OUR LAST ADVENTURE

by the same author
Intelligence Came First (editor)
Occult Chemistry Re-Evaluated
This Dynamic Universe (joint editor)
Vitamin B^{12}

OUR LAST ADVENTURE

A Commonsense Guide to Death and After

by

E. Lester Smith

THEOSOPHICAL PUBLISHING HOUSE LTD.
68 Great Russell St., London WC1B 3BU

Adyar, Madras
India

Wheaton, Illinois
U S A

ISBN 0-7229-5122-1

Typeset by Avonset, Midsomer Norton, Bath

Printed by Whitstable Litho Ltd., Whitstable, Kent

DEDICATION

TO ALL WHO ARE AFRAID OF DEATH

'Oh, I'm dead! How lovely!'

ACKNOWLEDGEMENTS

I am deeply grateful to friends who read the first typescript of this book, and in some instances passed it on to their friends in turn. They all encouraged me by approval of my work.

But there were some, including my wife Winifred, Diane Summers, Joan Powell, Ianthe Hoskins and other members of the Editorial board, and particularly the editor, Helen Gething, to whom I offer heartfelt thanks, for they made specific constructive criticisms. These led to extensive revision of the text and the writing of the Epilogue, without which the book would have been incomplete.

I am grateful to Celia Green for permission to quote from *Lucid Dreams* and *Out-of-the-Body Experiences*. Also to R. A. Moody, R. A. Monroe, Michael Whiteman and John Fuller, whose books provide valuable evidence for conscious existence outside the physical body.

My warm thanks also to Joan Thompson, who managed to decipher my handwriting, typed the first draft of the book, and took a personal interest in it.

CONTENTS

PROLOGUE

TRAVELLING ADVENTUROUSLY

Ever since my teens I have enjoyed planning adventure holidays. At first I took day trips alone, on foot or bicycle; later I joined a rambling and camping group, becoming the leader in due course. We took it in turns to organise Sunday rambles, travelling by train from London to some starting point in the Home Counties, taking a sandwich lunch and having tea at a convenient café. In the summer we sometimes arranged weekend camps, taking our own lightweight tents and preparing food over primus stoves. At Easter and Whitsun we usually arranged a camping tour walking, for example, round the Isle of Wight or along the South Downs. Then sometimes we planned summer holidays together in smaller groups, camping or staying in Youth Hostels or inns, either in the British Isles or in Europe, always carrying our minimal needs on our backs in rucksacks.

The point I want to make is that all these trips had to be *planned in advance* by the solo adventurer or the group leader. The planning had to be careful and thorough yet flexible. If it was haphazard the holiday might end in disaster because we had failed to reach the finishing point in time to get home and start work on the Monday; if it was too rigid, then the adventure was largely in the mind of the organiser and the trip became too regimented and less enjoyable. There was a need to allow for contingencies: for bad weather, for more rest if it became too hot, for taking detours or side-trips that looked interesting, making up the time by shortening later stages. All this introduced a sense of fun and adventure into the trip itself: any hour, any day, might bring an unexpected new experience to add to our enjoyment.

At that time most people's idea of a holiday was to return to the same boarding-house in the same seaside resort year after year. A limited sense of adventure might be satisfied by switching from, say, Southend to Margate. Inexpensive package holidays had not yet arrived and relatively few people travelled abroad. Our tours, on the other hand, though physically strenuous, stretched our youthful limbs and lungs and were immensely rewarding and memorable.

After graduating, I started work at an industrial research and development laboratory. A few years later, feeling the need for a break following a period of intensive research, I asked if I could take a week of my annual leave immediately. My request was granted — so what should I do with the time? Looking at a map of the British Isles and having a liking for islands, I decided to walk round the Isle of Man, which I had never visited. On the way home I bought a one-inch Ordnance Survey map of the island, collected what information I could obtain free from a travel agent and enquired the times of trains and boats. At home I put a minimal change of lightweight clothing in my rucksack along with other essentials and food for a few midday meals. As a vegetarian from birth, I was satisfied with hard biscuits, cheese, nuts, chocolate and fruit. My plan was to seek bed and breakfast at private houses; in the late 1920s about 2s.6d. a night was a reasonable charge. I would have my evening meal at a café or where I spent the night.

Next day I took a train to Liverpool and on the journey studied my map and planned my walking tour. I intended to follow the coast except for a few excursions inland. Should I go clockwise or anti-clockwise? The latter seemed preferable. Could I plan each day's walking to end at a town or village where I might hope to find accommodation? Could I walk the whole distance or would I need to travel partly by train or bus? The plan was made in outline only — there was scope for alterations as circumstances or my fancy suggested.

On the crossing I started to enjoy my holiday. The boat arrived in the evening at Douglas, capital of the Isle of Man and the resort that attracted most of the holidaymakers. I was in no mood for the bright lights — I wanted the solitude of mountain and cliff path. I leapt upon one of Douglas's famous trams — in those days small, open-sided, with benches across and horse-drawn. It took me northwards towards Laxey where I sought and found accommodation for my first night. Next morning I set out to climb Snaefell, the island's highest mountain at just over 2,000 feet, stopping on the way to admire the enormous Great Wheel of Laxey. This is not part of a fun-fair but an example of early industry — a pumping station to take water out of the lead mines. The weather was fine and bright and I enjoyed my steady climb up the mountain, but its top was wreathed in cloud and visibility was reduced to a few yards. Near the grassy summit I lost the track and wandered around bewildered. My compass told me that I must be walking round in circles quite near the top, so I turned up the now gentle incline and soon reached the summit. Most people would have felt lonely on that desolate mountain-top. Actually I did not — I was acutely aware of my alone-ness (which is not at all the same thing) but was rejoicing in it. Peering through the mist, to my amazement I saw another human being. I was not only astounded, but almost disconcerted, to meet the local shepherd, minding his flock on that lonely height. 'Ah,' he said, 'you should have been here yesterday; you could see five countries.' (Isle of Man, Ireland, Scotland, England, Wales.) I could not wait for the cloud to lift, so missed this famous view. I made my way down the other side of the mountain towards the north of the island, along narrow roads that form part of the route of the annual TT motorcycle races. I went on to visit Ramsey, 'the Queen of the North' and then across to the west coast.

The following day my route lay along the cliff tops of the island's west coast. There I gloried in the fierce wind and the sight of the wild sea below — these were the highlights which still stand out in my memory fifty years later. The rest of the tour went according to plan. I went on by Kirk Michael and the old town of Peel, then turned inland to see the Tynewald Hill. Once a year on Old Midsummer Day, July 5th, the House of Keys, the ancient Manx parliament, still meets here ceremonially in the open air. I continued over S. Barrule and the quaintly named Cronk-ny-Irree-Lhaa to the pleasant holiday resort of Port Erin, and on to Spanish Head to look across at the beautiful little island, the Calf of Man, in the south; thence east again to Port St. Mary and the picturesque Castletown, the old capital, with Castle Rushen and a delightful bay, and back to Douglas. I returned to home and work exhilarated and refreshed.

A year or so later I had a far more adventurous holiday in the Alps with three friends all a little older than myself, from our rambling group. We travelled by train and boat to Geneva, then on to Martigny. From there we began a walking and climbing tour that was to take us in a big ellipse, fifty miles long, all round Mont Blanc. This highest mountain in the Alps bestrides the place where three national frontiers meet, those separating France, Switzerland and Italy. Our trip was a good example of the need for plans to be highly flexible as well as thorough. We had no idea, when we started, of the dilemma that lay ahead; it was to be resolved by a dangerous and illegal crossing of one of these frontiers!

We started walking, uphill all the way, to the idyllic little mountain resort of Champex at 4,800 feet. As well as glorious views of mountain and forest and exciting climbs for day trips, it had a lake, just one mile in circumference, a perfect after dinner walk on a starry evening. One day we climbed to the Cabane d'Orny, a mountain refuge, at the foot of the Orny Glacier at 8,800 feet. This, we thought, would be the highest point of our holiday but, in fact, we were to climb over 2,000 feet higher later on. For the present, we stayed at Champex a few more days before walking mainly downhill to rejoin the mountain road from Martigny over the Great St. Bernard Pass. We reached the road at Orsiers which is a typical mountain village built from stout blocks of local stone, roofed in heavy slabs of slate against winter storms. Branching off to traverse the Val Ferret and climb over the Col Ferret, we came to the Great St Bernard Pass at 8,100 feet. There we visited the Hospice and saw those famous St. Bernard dogs that carry little casks of brandy to travellers trapped in the winter snow. Continuing on downhill we passed the picturesque village of St. Remy to Aosta in Italy and on to Courmayeur for a few days rest. There the blow fell.

Mussolini was flexing his muscles at this time and we heard the news that he had closed all the passes out of Italy, except those of the Great and Little St. Bernard. This was most inconvenient; we had come in by the Great St. Bernard, and the Little St. Bernard Pass would take us far from our intended route. So what should we do? Well, we consulted the local mountain guides and they told us there would be no difficulty if we followed their advice.

They promised to get us out of Italy, but were a bit cagey about how they would do it. Were we good climbers? Not really. Ah well, we should manage, but we would need ice-axes! Ice-axes! Here was adventure with a vengeance; we certainly did not regard ourselves as mountaineers. We might never again need ice-axes, so we did not want to spend our slender savings to buy one each. Could we perhaps hire them? Yes, that could be arranged. The next day we duly set off with our guide, carrying our belongings in rucksacks — plus an ice-axe each. We ascended through the wooded foothills until we emerged above the treeline. There we saw ahead of us a gigantic vertical wall of rock stretching into the far distance to right and left. It was clearly impassable, yet we were making straight for it! As we came closer, we saw there was a tortuous route winding up the rock-face like a jagged, uneven, interminable flight of steps with some of the rocky treads three or four feet high. Slowly, panting and gasping with the exertion in the thin air, we climbed and climbed to 11,000 feet — two miles above sea-level, until at long last we reached the top. This was more level ground and we saw ahead the mountaineers' hut, the Riforgio Torino where we would spend the night. But we were still on Italian soil; our defiance of Mussolini would come tomorrow.

At the refuge we had our supper, then went early to the loft to sleep rough. The guide insisted that we make a very early start, at 4 a.m., so that we could cross the snowfield before it was softened by the summer sun. Never before or since have I felt so cold, except perhaps when swimming in unexpectedly cold water. I was dressed lightly, in shorts, for hot summer days; but this was night, at 11,000 feet and with an icy wind blowing across miles of snow and ice.

As soon as we reached the snow, we were all roped together and we found a use for our ice-axes, as a sort of third leg, on the frozen slippery surface. Somewhere up there on that desolate snowfield we crossed the unmarked frontier into France. Sure enough, there were no Italian frontier guards up there to stop us. We had ascended over the Col du Géant, a shoulder or pass on the Mont Blanc range, two-thirds the height of the great mountain itself. As we descended on the other side the packed snow gave way gradually to hard ice. But it was far from level; the intense pressure of winter snows had forced it up into great hillocks and ridges. We needed to cut steps in the ice to negotiate some of these. But far worse, across our path from time to time lay huge crevasses, long deep splits in the ice, too long to go round, too wide for a standing jump, too dangerous for a running jump on the slippery ice! So what could we do? The tall athletic guide could and did jump first, and called us one by one to follow. He braced himself, held the rope just ahead of the jumper and gave a judicious tug when he or she was in mid-air. A rather short girl in the party was terrified, knowing for certain that she could not manage to clear the crevasse and would fall into the icy black water far below. But of course she did make it, every time, with the guide's help. If one of us had fallen in we could have been hauled out by rope, probably unhurt though shaken. Thus we made our way down that most famous glacier of them all, the Mer de Glace. When it became possible we crossed over to the left-hand edge of the glacier and so on to the jumbled boulders of the moraine which is formed from rocks great and small, torn off the mountain by the ice and spewed out on each side and at the lower end of the glacier.

So we scrambled and stumbled on, descending nearly 5,000 feet to Montenvers near the foot of the glacier. Here there was a restaurant, where we had some needed food and drink, and congratulated ourselves. What intrepid mountaineers we were after all! We had climbed the mighty Col du Géant, two-thirds the height of Mont Blanc itself, and we had traversed the whole length of the Mer de Glace! But then our guide re-appeared, to bid us farewell — and took all the wind out of our sails. Shouldering our four hired ice-axes, he proposed to climb back alone up the glacier, and on down that terrible rock-face, right back to Courmayeur in what was left of that day. We felt truly deflated, but still we *had* made it, with his help, and for all of us it has remained the thrill of a lifetime. Now, it is possible to make the whole crossing in an hour or so, by a succession of cable-cars. Looking back in the light of later experience I know that the rock-face was not dangerous, just horribly tiring; but the glacier would have presented deadly hazards if we had tackled it without a guide. Our greatest danger was the possibility of a sprained ankle, because we were wearing only stout shoes instead of the climbing boots that some of us bought later for other expeditions.

Scorning the rack-and-pinion railway, we walked down the zigzag road to Chamonix. From there we walked on, still in France, to spend a few days at Argentiers before returning via Trient to Martigny, and so by train and boat back to England.

Fortunately the spirit of adventure still persists among many young people. Some still walk and camp, though if they walk, they are more likely to stay at Youth Hostels and if they camp, they are more likely to carry tents on motor bikes or in cars. Nonetheless, I wonder if they get the thrills we had doing it all the hard way on a shoestring. In later life I too have arranged tours by car, or business trips, mostly by plane, around Europe and America, all very enjoyable and still needing thorough planning, but definitely less adventurous than those early holidays.

Today the spirit of adventure is also damped down by package holidays. It is so easy to pick up an armful of glossy brochures from a travel agent and simply select the most glamorous holiday available, at a price you can just about afford. With charter flights and cut-price hotel booking for parties, it works out cheaper than doing it on your own.

Why am I telling you all this? What connection has it with the subject of this book? Simply this. There is a journey ahead that none of us can avoid — our last adventure, death and beyond. If you travel abroad, it is wise to take a map and a guide to the country. Most people face death totally unprepared. Some are scared and refuse even to think or talk about it; others are resigned or willing to go when their time comes; very, very few have any clear idea of what lies ahead. I suggest this may be foolish and unnecessary, that it will be an easier, maybe even an enjoyable experience if you know something in advance about what to expect. How *can* you know? In the chapters which follow, I have tried to write a commonsense guide to the next world.

Do not push my analogy of overseas travel too hard. You are not really going to a new *place*, some sort of heaven in the sky. You will be moving into a kind of inner space which *you know already* to some extent. Many people have written of this inner space. Let us explore this space in imagination, in a commonsense, unsentimental fashion, drawing upon and enlarging our own experiences. This book is the result of my own reading and thinking, enhanced by periods of inner quietness. As I have grown older I wanted to experience this adventure in imagination. Please come with me for company.

CHAPTER I

LIVING OUTSIDE THE BODY

This book is written mainly for people of the Western world, most of whom profess allegiance to one branch or another of the Christian faith. Some do so sincerely, others rather nominally, while some are frankly agnostic. So it may be necessary to start by trying to convince you that conscious life is possible at all outside the physical body. This is a sad state of affairs. If I were writing for other parts of the world where different religions are widely followed, it would hardly be necessary. For the facts of life after death are taught, in outline, in such religions, and the faithful believe these things. If you are a sincere Christian you may say that this is taught also in Christianity. True, but is it taught effectively? Is it believed? It forms part of the burial service but otherwise the topic is usually avoided both by the Church and the worshippers. What information is available these days is scanty and largely misleading, especially in Protestant churches, and is seldom spoken of. There is an embarrassed conspiracy of silence.

Fortunately, I do not need to ask you to take what I have to say on trust. There is plenty of evidence available and it has been analysed in a scientific manner. But it is not widely known and it cannot be claimed that the evidence is accepted by the majority of scientists. I think this is a pity, because I am one myself and I know that in these days people tend to pay more attention to the words of scientists than to those of ministers of religion. But most scientists are not interested because such ideas do not accord well with their current beliefs, so the ideas are simply thrust aside without study. Scientists in general are not such open-minded seekers after truth as is generally believed. But that is another story.

I might start by asking what you suppose happens when you go to sleep. What we call the waking consciousness is then in abeyance, but can you believe that you, your true self, have become unconscious? That would mean that you virtually ceased to exist while you slept and that you are re-born, so to speak, on waking. I could speak of dreams as some sort of evidence for activity apart from the body. But I shall not do so; there is less shaky ground upon which to base my case. Dreams are usually incoherent, meaningless fantasies, and it is easy to imagine that they are simply conjured up by the brain from old and recent memories when it is freed from conscious control.

That leads me straight to another argument. If you feel, as I do, that the brain is indeed under control during waking life, then who or what is controlling it? *You* are, of course — but who are you? Used in that sense, 'you' surely has to mean some *non*-material being, some entity additional to the physical body? For some of you — I hope for most of you — I am merely stating the obvious. If you feel that is so, then I must tell you that many materialistic scientists sincerely believe that body and brain suffice to explain the whole of man. Back in the last century, the idea was crudely expressed in the words, 'the brain secretes thought just as the liver secretes bile'. In these days it is put more delicately, but the meaning is much the same; mind and brain are simply different aspects of one and the same thing, these scientists proclaim. Moreover, since our civilization is based so successfully upon science and technology, many thinking people believe that scientists must be right, so they adopt their ideas.

Wilder Penfield, a distinguished American neurosurgeon, started out with just such beliefs. His career involved a great deal of pioneering research on the brain along with brain surgery for the relief of epilepsy and other conditions resulting from brain damage. His own experience in the course of this work compelled him, unwillingly at first, to realise that mind is a separate non-material entity that directs and uses the brain and body for its own purposes. Likening the brain to a computer, he concludes that the mind is then analogous to the programmer, to the intelligent human being who manipulates the electronic computer as a tool for rapid calculations. When he retired, Penfield wrote a splendid little book, *The Mysteries of the Mind**, in which he charts the many episodes that led to his 'conversion'.

Sir John Eccles, an even more distinguished British scientist, who spent a lifetime in brain research, had just the same experience. Again, late in his life, he wrote a magnificent book jointly with the philosopher Sir Karl Popper: *The Self and its Brain.* Near the end he wrote, 'So I am constrained to believe that there is what we might call a supernatural origin of my unique self-conscious mind or my unique self-hood or soul. . . By this kind of supernatural creation, I escape from the incredible improbability that the uniqueness of my own self is genetically determined.' If you want to follow the thinking of scientists, then follow men such as these. Each was in the most favourable situation possible to discover the truth of the mind-brain problem, and each had the courage to formulate fresh conclusions and to admit that he had been wrong at first.

Earlier, I declined to derive any support from dreams for the idea of a superphysical body. But there are two special kinds of dream that some people have occasionally, or even regularly, that do offer some evidence. The first is a vivid dream of meeting one's mother or another close relative or friend who has died. Almost always this dead person appears joyous, well and fit again, as in the prime of life. Sometimes a message or conversation is recalled on waking, more often just a sense of happy communion. I believe these represent 'dreaming true' and an actual meeting in the other world between the sleeper in his subtle body and the 'dead' person in his or hers.

The other type is called a lucid dream. The word 'lucid' does not mean that the dream is especially clear and bright; it means that the dreamer *knows,* during the dream, that he *is* dreaming; in other words the lucidity is not visual, but intellectual. In lucid dreams you think and act much as you do in waking life; you have some control over the course of the dream, though usually not full control. Such dreams are sufficiently well known to have brought forth at least one book on the subject, *Lucid Dreams* by Celia Green, of the Institute of Psychophysical Research. It is based on case material available to this Institute plus additional material

* Books mentioned in the text are listed at the end of this book.

from books on wider topics. In this kind of dream you are usually aware of yourself in some kind of body, which is clearly not the ordinary solid, heavy physical body. For example, it is able to float or fly without effort! Jumping and flying are common features of lucid dreams as illustrated by the two examples which follow. The first is from Celia Green's book:

> At one time I had lucid dreams very frequently. I found it very interesting to do things which I could not have done in normal life. For example, I used to go into shops and wonder if it was a real shop, and whether anyone would notice if I took something. I used to enjoy jumping from a height in these dreams. I might be in my flat, or in some other building, but it was quite safe and very pleasant to jump from the window. However, eventually I became alarmed and gave up the practice because I was afraid I would begin to think I was dreaming when I was not, and then I might try to jump from a real window.

In the second, recorded by the Russian writer, Ouspensky, it appears that nearly all the normal senses can appear to function in these dreams.

> I hear their measured tread and watch them from the window of a high house in Galata, in Constantinople, in a narrow lane, one end of which leads to the old wharf and the Golden Horn with its ships and steamers and the minarets of Stamboul behind them. The Roman soldiers march on and on in close ranks along the lane. I hear their heavy measured tread, and see the sun shining on their helmets. Then suddenly I detach myself from the window-sill on which I am lying, and in the same reclining position fly slowly over the lane, over the houses, and then over the Golden Horn in the direction of Stamboul. I smell the sea, feel the wind, the warm sun.

Unlike normal dreams, lucid dreams are usually remembered clearly on waking; so you need not be afraid of having one and forgetting it. If you should have one of these lucid dreams, then be careful not to get too emotionally involved with it, or the dream is likely to end immediately. Emotional conflict is particularly to be avoided. The final example from Oliver Fox concerns this, and also mentions the sense of taste, the only sense not included in Ouspensky's account.

> It was so difficult to maintain the role of an impersonal observer in this strange Dream World, to realise that if I allowed my emotions to get the better of my mental control the dream would come to an abrupt end. I would enter a restaurant and order a meal, only to wake after savouring the first few mouthfuls. . . I would encounter a fascinating lady and even

talk to her for a little while, but the mere thought of a possible embrace was fatal.

The lucid dream merges into a slightly different and better-known kind of experience. This is the 'out-of-the-body experience', or as others have named it, 'astral projection', or more simply 'separation'. It is likely to start from the waking state, with scarcely any break in consciousness; it is as if you fall asleep, yet remain 'awake' in a different body, in a different world or a different kind of space. However, such experiences can start from normal sleep, or from unconsciousness caused by anaesthetics given for an operation, or from concussion caused by a fall or other accident. In this state of separation you are likely to be more fully aware of yourself in the immaterial body than in a lucid dream, and quite often you are able to look at your own physical body from outside. In less common cases it is not even necessary for this ordinary body to be unconscious; it may continue with whatever it was doing before separation occurred — talking, singing, typing, walking, even driving a car or motor cycle! But you, with your sense of I-ness, the real you, are always in the separated body. You can watch your normal body still talking sensibly or going about its business as if it belonged to someone else. You may regard it just as objectively, criticising its manner of speaking or what it is saying, for example. For the duration of the experience you have no special interest in this ordinary body and no personal connection in consciousness with whatever it is doing; you are happily and vividly conscious in your separated body — except in some rare instances of dual consciousness. Of course, there must *be* a connection, or the body would sleep or faint, or behave noticeably as a mere automaton; the point is that you are not usually *aware* of the connection, you seem to be completely outside. However, it is far more common for separation to start when you are in bed, or resting in a chair.

I should like to be able to write convincingly of my own experiences in separation, but they are too slight to be worth recording. There are far better examples to be found in the numerous books that cover this topic. So, to *you*, my own experiences are unimportant. But to *me*, they are crucial. They enable me to write from vivid personal experience, altogether beyond argument. It is not a matter of belief. I *know*, without possibility of doubt that I do have this second super-physical body; and because I do not suppose that I am unique, I fully believe that you and everyone else has such a body, even if you are not yet aware of it. But, in fact, you probably *are* aware of it already, even if not in this dramatic manner. Most civilised people do not live *in* their physical bodies to any great extent. Their interests are centred at most times at higher levels; they tend to live in their emotions and thoughts. If you reflect on this, you will realise that when you are in the grip of emotion, or when something serious compels you to think hard, then at these moments you are scarcely aware of your body. You are conscious in some other realm, in some other state of being, effectively in a non-physical subtle body. I shall return to this line of reasoning in a later chapter.

Meanwhile, more dramatic evidence comes from people's reports of separation or out-of-the-body experience. I have mentioned Celia Green's book *Lucid Dreams*. She has written another called *Out-of-the-Body Experiences* which analyses these strange happenings from every possible angle. It is an interesting little book which, unlike the other, does not draw at all on previously published accounts. She arranged for an appeal to be published in the press and broadcast on radio, asking anyone who had such experiences to write to her. The hundreds of people who did write each received a questionnaire to fill in. 326 replies were received and these people then got a second questionnaire, to which 251 of them responded. The book is based upon their experiences. So here were 300 or so ordinary people who not only had these odd experiences but happened also to see or hear the appeal and who were interested enough to fill in questionnaires. So this must be just a small sample: there are many thousands of people among us who have known separation at some time but who seldom mention it for fear of seeming odd or different. As I shall mention later, another author managed to collect over 800 cases. So these happenings are not so very unusual after all.

The first thing that must be said about this state of separation is that it is nearly always reported as a condition of delightful freedom and happiness. Many people accept it immediately and revel in the experience. Others, it must be admitted, are puzzled and scared by its strangeness; they may even wonder if they have died. Or after a while, having enjoyed the adventure, they feel suddenly lost and wonder how they will ever find their way back. I know — it happened to me. Actually, such fears are groundless; as soon as they are felt, the separated body is automatically drawn back to rejoin the physical, and the subject wakes up, sometimes with a jolt, remembering what has happened. After a few such experiences you know what to expect and that there is no cause for fear.

The next important thing to mention is that most of these people are aware of being in a subtle body, which usually closely resembles their ordinary physical body. It may seem younger, because it is free from any disabilities that may afflict the normal body, from old age, illness or accident. If it has lost a limb, for example, the subtle body is whole again. It usually appears dressed either in the clothes or night clothes actually being worn at the time or in some other garments habitually worn. Some people see themselves just in vague filmy wrappings, usually white. This is the way ghosts often appear. Is that so strange? Does it occur to you that in this separated state you *are* a ghost? If you succeed in making yourself visible to some other living person (and that is not often possible) he or she will almost certainly regard you as a ghost, and may wonder if you are dead. Occasionally the apparition seems so real that it is mistaken for an actual person in his physical body — until it disappears or passes through a wall and out of sight. Another common feature is that many subjects look back and see their own body lying on the bed, sitting in a chair or occasionally awake and engaged in some activity. Their usual reaction is not alarm, shock or amazement, but a rather casual attitude. 'Oh, that's me; so that's what I look like to other people.' What does seem surprising is the lack of interest felt by many subjects for their inert physical body, their normal abode. The reason is that their consciousness is almost completely centred in their separated body; it is *this* which feels vividly alive and the other body is scarcely more interesting than an overcoat discarded on coming into a warm house.

> I had no further interest in my physical body, or indeed my physical life. I only wanted to pursue and prolong this happy state of being where everything was more bright, vivid and real than anything I had previously known.
> Celia Green *Out-of-the-Body Experiences*

For those who have experienced it, this ability to look back and forth from one body to the other is irrefutable proof that they do indeed have this second body in which life can continue quite independently of the physical frame. Subjects emphasised the 'positive' quality of the detachment they experienced, not resignation. Some felt great interest in what they experienced but no emotional involvement; emotion is liable to lead to return. Interest was described as 'academic', 'critical' or 'objective interest and fascination'. 'Problems of day-to-day life are very trivial.' 'It was as if all the worries of my life had left me.' But some subjects felt pity for their suffering bodies.

Thinking and memory persist in the separated form, often more acutely than in waking life. Thinking of another place is enough to send one floating there almost instantaneously. Separation is sometimes caused by stress, such as the realisation of impending danger, as illustrated by the following case; it occurred before the subject actually hit the ground in his fall.

> My experience took place at 'Harrison's Rocks', a small rock-climbing cliff south of London. After sleeping out for the night I rose on a Sunday morning and (before eating anything, as I recollect) began to 'solo' the rock climbs there. About 15 or 20 feet above the ground, I slipped and fell. . . Objectively, the height wasn't great, but I think that I was very frightened of the coming pain or death, and for a moment I abandoned the idea of living. As I fell, I seemed to be about 5 or 10 feet out from the rock face, looking at my body falling (in contact with the face). I vaguely recollect wondering if I could investigate this odd sensation by moving round to the other side of my body to look at it. Once I hit the ground, I was immediately preoccupied with my pain: I had badly sprained both ankles. . .
>
> Celia Green *Out-of-the-Body Experiences*

Describing the sense perceptions of the subtle body poses problems. Obviously in separation it cannot use the sense organs of the physical body; nor does it have a duplicate set of its own, though one might think so after reading some accounts of separative experiences. Only the more experienced observers recognise that the subtle body becomes aware of its surroundings in a totally different and unique fashion. It is as if it had a single sense organ, which combines the capabilities of all five normal senses — and more besides. Perhaps there is no localised sense organ at all, but the whole body is sensitive and receptive. But how can this be described? We have no words for it. The term *synaesthesia* has been used, meaning a synthesis of all the senses. But it is not very useful in ordinary accounts of experiences. So what almost all writers do, even those who fully understand the situation, is to use the familiar words: seeing, hearing, feeling, etc. But in reading such accounts we should imagine these words to be in inverted commas; they do not have their ordinary meanings. They are used in an 'as if' manner. So if someone says 'I was out of my body and I saw an Angel', you should mentally rephrase the statement somewhat as follows: 'I was out of my physical body, and in my subtle body I became aware of a Being. I cannot properly describe what happened. It was as if I saw him, but it was much more than seeing as we know it on earth. I also knew that he was a loving spiritual Being of great splendour, and that he had come to help me. He reminded me somewhat of the descriptions of Angels that I had read, but he was altogether more glorious, more loving, than anything I had imagined.'

It is largely this manner of awareness that makes life 'on the other side' so mystifying at first — but also so much more vivid, more satisfying than our earthly ways. By comparison, these seem primitive, notwithstanding the truly marvellous capabilities of eyes and ears, in combination with the brain. Other capabilities of the 'disembodied' state are the ability to 'see' in every direction at once, not just from the position the eyes would occupy, and to see both sides of an object simultaneously. This accounts for the occasional reversal of an object or number on return to the body. Thus the number '78' may be remembered as '87' because of the difficulty of distinguishing right from left. Also it is possible to see in the dark when out of the body, and to see through objects, or to see them outside and inside at the same time. The sense of time may also be different from that of the normal waking state.

Several people have tried to visit a friend when out of the body. Usually the friend is not aware of the visitor but sometimes the presence is either seen or felt as an impression.

Psychokinesis, i.e. moving objects psychically by thought or by the separated body, is not usually possible. The subject is often aware of his subtle hand just going through the object he wants to move without affecting it in any way. Similarly the separated form can go through a ceiling or wall without hindrance; but often the subject is inhibited by his normal experience and does not try such experiments.

Some fifty years ago or more, several writers sought to explain the puzzling nature of the astral plane* by suggesting that it is a world of four dimensions instead of the three dimensions of the normal world. To help us to visualise the situation an analogy was presented. We were bidden to imagine a race of flat-men having no thickness and confined to a flat two-dimensional (2D) world. To them our normal three-dimensional (3D) world would be as mystifying as a four-dimensional (4D) world would be to us. The analogy was not very convincing however, because it is almost harder to envisage 2D men than 4D men. I think the reason is that we are surrounded by representations of the 3D world on flat 2D surfaces, we have drawings, paintings, photographs and now television pictures. With the help of perspective and the co-operation of eyes, brain and mind we have no difficulty in filling in the missing dimension of depth by imagination.

There is, I think, a better analogy, that of the shadow-picture or silhouette. If a person's head in profile, for example, is lit by a single concentrated light, it can throw a sharp shadow onto a white wall. There you have a truly 2 D projection of that person's head. If he or she were unknown to you there would be great difficulty in imagining how that person would look full-face, for the shadow-picture gives no depth clues as a drawing does. This analogy does, I think, enable us to appreciate better our difficulty in trying to imagine the still greater depth, completeness and all-round vision of astral sight. Ordinarily we perceive from a fixed point, that of our eyes, and all we see of a solid object is the part of its outer surface visible from that viewpoint.

I am not suggesting that the next world is really four-dimensional in a literal (geometrical) sense. There is an alternative explanation of its properties, which I find more satisfactory. But the idea of a figurative fourth dimension may be helpful to some.

Other books on this subject include further compilations of case reports, notably by Crookall, and other books covering the authors' own repeated excusions into this other world. Perhaps

* The astral plane was considered to be the place where the subtle body existed. In modern writing the word 'level' is used rather than 'plane'.

the best known of these is an early one first published in 1929, by Muldoon and Carrington *The Projection of the Astral Body.* Muldoon lived in a rural part of the United States and he experienced what he calls astral projections from the age of 12. He was only 25 when he started writing his book, encouraged by the other author, a well-known writer on occult matters. By this time he had already known hundreds of these experiences, some spontaneous, some deliberately induced. Separation came easily to him and he writes in the conviction that most people could experience it for themselves by following his technique; in practice, it is by no means so easy as he supposes. He does admit that it may be difficult for people in robust health. Surprisingly, he claims that separation is easier when the body is fatigued, ill or aged: he induced a suitable state of passivity by slowing his heart-beats — something he could do at will, but not a very common achievement. Then relaxation of the body is recommended, while keeping the mind alert, thinking about projection and strongly imagining the astral body rising above the recumbent physical form in bed; but the emotional state must be calm and without fear. For Muldoon, however, projection in full consciousness was sometimes an unpleasant experience, accompanied by sensations of up-and-down vibrations and pain at the back of the head.

Other practitioners, notably Monroe (see later) have mentioned loud strange noises in the ears, including popping noises, roarings and sounds like rushing wind. It is far preferable, this author says, just to fall asleep normally, and then 'wake up' into astral consciousness. Then one may wake fully, direct from this state, or may lapse back into ordinary sleep first. The direct awakening from astral consciousness, especially if it is due to fear, can be unpleasant and even painful, as the bodies come suddenly 'into coincidence', as Muldoon expresses it. The result is a 'repercussion' which jolts or shakes the body. I myself often get mild repercussions when trying to get to sleep while my mind is still active; I suppose that I slip away and immediately get pulled back to complete my train of thought. Anyhow, this notion of passage from sleep into astral consciousness has a familiar ring about it: this is what many of us do at night — only we seldom recall our excursions on waking. Actually, separation itself is nothing unusual; it is remembering that is the uncommon trick we need to learn. Muldoon affirms, from observations on his family and friends while in separation, that the subtle body normally separates from the physical during sleep. But generally it hovers about a foot above the sleeping body, itself also in a state resembling sleep, while both bodies get recharged, so to speak, for the work of the morrow. Sometimes the subtle body may wander about, wrapped in its own thoughts and taking little notice of its surroundings. But other people are fully 'awake' on the astral plane, yet rarely remember their travels in the morning. This is perhaps Nature's protective mechanism: in waking consciousness we have enough to worry about, without the burden of our night-time activities to mull over as well. Such people have, as it were, dual Jekyll and Hyde personalities — but it is likely to be Dr. Jekyll who is active at night on altruistic missions. I shall return to this topic later.

Many writers have described a 'silver cord' connecting the physical and subtle bodies. Muldoon was always conscious of the 'silver cord' uniting them. Not only could he see it, he could also feel it as a seemingly soft and elastic, but solid, cable. He also felt a sort of elastic pull from the cord, tending to draw him back to the body, and its tension seemed to increase until he got some 10-20 feet away from his body; but once he managed to go beyond this critical distance, the cord thinned out to a thread and ceased to pull him back. Nevertheless, when the time came to end the separation the cord always did draw him back safely. So, if you ever do have such an experience, try to remember that it is impossible to get lost; the cord will always bring you home. Only at death, we are told, is the cord finally severed.

Muldoon's numerous experiences were mostly of a rather humdrum kind; he found himself in a world that closely resembled his normal surroundings, in a body like his physical body except that it was light and could travel freely. One of his most interesting and convincing experiences was when he visited a strange house while in astral projection. He was able to note its situation, in a nearby town, and to memorise details of its interior. He saw the family who lived there including a girl of around his own age who was sewing. Later, while awake and walking, he met and recognised this girl, introduced himself, and described details of her home which she confirmed as correct. They became friends and made some pre-arranged experiments in astral projection.

Another author who has reported upon his own numerous out-of-the-body experiences is Monroe, an American electronics engineer and business man. He suddenly started having these experiences, which continued and became more interesting. Being a science-orientated person, he made himself available for medical checks and scientific experiments and recorded his observations more scientifically than most other subjects have done. His experiences in what he calls Locale I, which closely resembles the ordinary world, were indeed similar to those of other subjects. But in his Locale II he found a less familiar world in which he had many adventures, met interesting people and 'conversed' with them, usually by a kind of telepathy without speaking. Time and place seemed to behave differently; at his slightest wish he found himself in a new locality.

Besides writing a book, *Journeys out of the Body,* Monroe followed up his own experiences by devising techniques for inducing the same condition in volunteers. He was aware of vibratory sensations in his body and sounds in his ears, including loud roaring just before leaving the body. So he got volunteers to relax on a water-bed, wearing stereophonic earphones, and fed various sounds into them. A pure tone in one ear and a slightly different one in the other were combined in the brain to produce a low-pitched vibratory beat tone. Adding 'white noise' at high volume to simulate the roaring sound, plus temple bells and other more musical sounds, was enough to drive most subjects out of their bodies. They retained a measure of dual consciousness, however, and could dictate their experiences into a tape-recorder as they happened. When I last heard he had succeeded with 600 subjects but much of their material awaited analysis. Even more interestingly, Monroe has formed his best subjects into an Explorer Group, whose members go out repeatedly to examine these other worlds. Moreover, when one of them is 'out' and reports a strange experience, a second member of the team can also be sent 'out' to find him and to confirm his observations. The fact that experiences can be shared is particularly significant and means that they are much less likely to be hallucinations of some kind. Previously I had been a little bothered that so many subjects seemed to have their own unique experiences. But this is not really surprising; if a hundred visitors from outer space were set down at well-separated places on earth, they would tell a hundred quite different stories of their adventures. There is no reason to suppose that the other worlds are more uniform than our familiar one.

Whiteman's book, *The Mystical Life,* also draws largely upon his own spontaneous experiences numbering several thousand. He is a university professor and he, too, has subjected his experiences to scientific analyses and cataloguing. He describes a number of differing types of separation and concludes that the grade of separation depends upon the degree of conquest over emotional and intellectual fixations. In separation he was not limited to earth-like realms or spaces, like most people; he describes higher spaces of increasing purity and spirituality. Whiteman's loftiest experiences were comparable with those of the famous Christian mystics and of the most experienced Indian yogis. These are more appropriately deferred to a later chapter; however, Whiteman's spiritual attitude to his experiences should be mentioned. They were all spontaneous; he never tried to induce separations though he did sometimes try to modify them, seeking to raise his consciousness to higher spiritual levels, through Love and Obedience. He regarded each event as being granted to him to enhance his understanding; he has repaid these gifts by writing books that may lead others towards the spiritual life.

Returning to the compilers of other people's experiences out of the body, I must mention that three books by Crookall together refer to no less than 800 cases. With some exceptions which belong to the next chapter, they do not add much to the information I have compiled from other sources. But what did interest me was the title of his last book, *What Happens When We Die.* As I shall try to show, this is a presumptuous and misleading title. To my way of thinking, it deals only with what may be presumed to happen to some people during the first few minutes or hours after 'awakening' on the other side. However, this book title does raise an important question: may we assume that experiences out-of-the-body during physical life accurately foreshadow the life after death? Just reading about these experiences may not be convincing; it is possible to dismiss them all as dreams or hallucinations, despite the 2,000 or so recorded cases and the many thousands more that must have been guarded as precious private memories. But even a single vivid *personal* experience of living in a separate body, and especially so if the physical body is also seen, *is* utterly convincing. That is to say, it provides irrefutable evidence by direct experience, that it is possible to function in such a 'Body of Light', as it may be called. If he thinks about it at all, such a person is likely to reflect: If I have such another body — and I *know* now that I have — then when I die, what is to prevent me carrying on living consciously in this body? What indeed? It is difficult to put down an argument like that. Personally I agree with it, except that my reading leads me to expect that there may be differences in detail, in the quality of life after death in comparison with brief episodes out-of-the-body. In particular I believe that as we become more familiar with after-death conditions, life opens out and blossoms in a manner that few have experienced during their life on earth. If you are still unconvinced please read at least the next chapter which offers more direct evidence.

CHAPTER II

BACK FROM THE DEAD

In the previous chapter we realised that most people who have had an out-of-the-body experience feel quite certain that they will encounter similar experiences after death. 'How could it be otherwise?' they will say. 'I *know* from my own incontrovertible experience that I can be vividly alive and conscious in this second body, quite separate from my usual body; so when I lose that one at 'death', nothing can prevent me living on in my second body.'

Nothing will shake their certainty. But the materialist who has not had the experience himself and does not want to admit such inconvenient ideas, that would upset all his previous convictions — what will he say? He will say it was a dream, or some sort of hallucination, anything to get out of accepting ideas to which his mind is closed. I doubt if we are *ever* going to get cast-iron *proof* of life after death — proof that even the most hard-headed scientist is bound to accept. All we can hope for is a vast accumulation of cases which, by their similarity and sincerity, may suffice to convince people with open, enquiring minds.

So how would it be if someone actually died and then came back, remembering what it was like to be dead? Thanks to medical progress in resuscitation, there are indeed hundreds of people who have done just that. People have died from drowning or from heart failure; sometimes resuscitation has been tried but has failed; in some cases even a death certificate has been made out. Yet something or other, perhaps the pleas of a relative, has made the doctor try once more, and the clinically dead patient has been brought back to life, by heart massage and artificial respiration, or with more modern aids. Some people to whom this has happened remember absolutely nothing of what they may have experienced on the other side. But others, hundreds of them, *do* recall these events and may relate them to the doctor or to a close relative. A few subjects 'died' twice with an interval of a few years. On one occasion they remembered nothing; on the other they had vivid experiences out-of-the-body.

Two American physicians, Dr. Raymond Moody, Jr., and Dr. Elisabeth Kubler-Ross, spent years collating records of such cases, quite unknown to each other. But it was Moody who published first under the title:- *Life After Life.* His publisher must have heard of the other investigation and sent an advance copy of the book to Dr. Kubler-Ross. She replied in effect that she could have written an exactly similar book herself because their findings tallied perfectly; she did, in fact, provide a foreword. The case reports are all given anonymously; in some instances that was made a condition by the subject.

The reports from these near-death subjects are similar in many respects to those we have been looking at from more or less healthy subjects. But they often include *additional* features, three of which are quite striking. More often than not the experience starts by the subject, released from his dying body, passing through a dark tunnel, with a light at the far end. On emerging, he is usually met by a benign, loving entity, who may be named as Christ or God, but who is often formless; it may be described as a ball of light simply, or as a Being of Light. This entity kindly but firmly invites the individual to review the whole of his past life. This is facilitated by a display resembling a kind of three-dimensional cinema or television screen, on which the major events of the life are shown; some subjects claim that every single episode, however trivial, was re-enacted. It is interesting that the Being merely directs the attention of the viewer, inviting him to reconsider his actions, especially asking how much true unselfish love each included. The Being does *not* pronounce judgement; he merely invites the person to judge himself. In the clearer vision and understanding of this other world he finds all his illusions and pretences stripped away, and is obliged to make a true judgement, however painful or humiliating this may be.

It might be thought that such a detailed review should take almost another lifetime, or at least many hours. But those who have experienced it often comment that time seems quite different in that other world — almost non-existent in fact, as though past, present and future were coalesced into the present moment. Frequently there is evidence from the progress of the resuscitation that the review can only have taken minutes of earth time, sometimes only seconds.

Another interesting point is that such an out-of-the-body event can happen to someone who was exposed to extreme danger, but who in fact escaped alive, with or without injuries. He was 'scared to death' and suffered a kind of psychological death in which he left his body briefly. An example was given in the last chapter of a climber who fell off a rock-face and left his body for the time it took to fall twenty feet or so. Moody records a case of a motorist who skidded off the road and who left his body during the time it took the car to fly through the air and land at the bottom of the embankment. He escaped with minor injuries. A further instance, taken from Moody's second book, *Reflections on Life After Life,* concerns someone who was trapped in a bad fire with explosions. He saw rescuers coming towards him but felt sure he would die before they reached him. In his terror he left his body and experienced a review of his past life. Then he returned and was astounded to see the rescuers hardly a pace nearer. He was led to safety though badly burned.

Moody does not record any complete case records. Instead he has analysed the experiences recorded by his subjects into 15 distinct features, each of which he illustrates with several appropriate snippets taken from his records. He emphasises that any particular subject will mention only some of these features; the others he may have experienced but failed to notice or to remember, or they may have been absent altogether. Moody appeared never to prompt his subjects: he just let them tell their stories and took notes or recordings. What he *has* provided, however, is a sort of idealised imaginary composite-case that does illustrate all 15 features. In practice no one person ever mentioned them all; the highest number reported by a few subjects was 12, though many reported 8 or more. Each element has featured in at least several accounts and some are nearly universal. The order in which they occurred, or were reported, differed somewhat from case to case.

It seems worthwhile to go through Moody's thoughtful analysis of 15 components of the experience. This will enable us to get a better feel of conditions immediately after death. Also it is helpful to note how closely his picture tallies with earlier descriptions: because, contrary to what he supposed, Moody was by no means the first to write books on such out-of-the-body experiences. I shall interpolate some of my own comments.

The first is *'Ineffability'*. However brief his experience, it becomes obvious to the subject that conditions in this other world are different in many respects from those on earth, even apart from sudden absence of pain and the feeling of lightness. There are just no words to describe that world; everyone agrees that this is so, but while some remain baffled, the more articulate individuals do attempt the impossible task of finding words, however inadequate, to describe the events. One woman explained that the physical world is three-dimensional; it has three fundamentally different directions at right angles — left-right, forward-backward, and up-down. But this other world has more than three dimensions. Now in the earlier decades of this century it was popular to describe the 'astral plane' of the spiritualists in terms of a fourth dimension (geometrical, *not* time as in Einstein's space-time world). I do not myself believe that this idea of a fourth geometrical dimension is literally true but it does serve to describe by analogy something of the peculiarity of this new kind of space.

Next comes *'Hearing'*. Immediately after release from the ailing physical body, the subject's attention is often directed back to the physical world he has so recently and mysteriously left. He may see his body and doctors or relatives trying to revive it; he may hear them saying that he is dead, and discussing how to attempt resuscitation. Doctors are sometimes astonished when they are afterwards told exactly what was said or done, by a subject who was certainly unconscious or even clinically dead at the time.

Most subjects report *'Feelings of Peace and Quiet'*. As one woman victim of a heart attack put it 'I began to experience the most wonderful feelings. I couldn't feel a thing in the world except peace, comfort and ease — just quietness. I felt that all my troubles were gone and I thought to myself, "Well, how quiet and peaceful and I don't hurt at all" '.

The next component mentioned by Moody is *'The Noise'*, though from other accounts, notably by Monroe, it is to be expected *before* the actual release occurs. The sound heard, by some subjects only, may be disturbingly unpleasant, or tolerable. It has been described as a loud buzzing or ringing; alternatively roaring, or whistling like the wind. Others may hear bells or beautiful music but this is more likely to have been heard while actually in the out-of-the-body state.

'The Dark Tunnel' experience has already been mentioned. Alternative descriptions are a cave or well, a funnel, a vacuum or void, and a sewer or a valley - the 'valley of the shadow of death'? A few people speak of falling or tumbling through space.

Moody devotes 20 pages to *'Out of the Body'* giving numerous excerpts from his case notes and comments on them. Since most people identify themselves with the physical body for much of the time, it comes as a shock to find themselves separate from it in another state of existence, from which they can sometimes look down and see their body. Not surprisingly they are often confused at first and may not link the experience with death. Some are afraid and may make panic attempts to get back into the physical state; others, as noted, are quite tranquil and immediately accept the new condition. We may suspect that they have frequently experienced it in sleep, though they have never previously brought through any memories to the waking consciousness. Such individuals are inclined to have no regrets at leaving the body and would happily go forward into the after-death life if they were not pulled back. One woman after an interval of confusion, remembered thinking, 'Oh, I'm dead! How lovely!'

A few people feel themselves to be pure disembodied consciousness, without any body at all. Others feel (or imagine?) themselves to have a body that in some ways resembles the physical one. Others are so taken up by the strange events that they do not even consider their own situation. The *'Spiritual Body'* as Moody calls it, is sometimes recognised first through its limitations; in this body it is impossible to communicate with other people who are alive and awake. To them one is invisible; they do not hear if they are spoken to; attempts to touch them evoke no response in them, nor in oneself any sense of contact. These living people may even appear to walk right through one's spiritual body, without noticing! Most frustrating!

The positive characteristics of the spiritual body are difficult to convey in ordinary words. There is general agreeement that it is weightless and lacks also the sense of equilibrium that keeps the physical body upright and steady it its movements. On the other hand, the spiritual body is free to move anywhere at will almost instantaneously and to pass through physical obstacles. When this body is perceived, it may appear transparent or cloud-like. The seeming timelessness of this state has already been mentioned. Another characteristic noted by some is an ability to think clearly and quickly; in other words one is using the non-physical mind, free from the slower-acting physical brain. Perception in this state does not seem to depend upon specialised organs like eyes and ears: it is as if the whole body could receive impressions, so that one can 'see' behind as well as in front, for example. Hearing may be translated into words, out of habit, but really it is more like telepathy — a mind-to-mind communication directly in thought, in ideas not verbalised. Another comforting observation, for any who are lame or have lost the use of some parts of their body is that, nevertheless, the spiritual body appears whole and perfect.

At first, inability to make any contact with living people may cause loneliness and a sense of isolation: but this is later dispelled on meeting other people in their spiritual bodies, including relatives who died previously.

This brings us to Moody's next section, *'Meeting Others'*. Some people in this near-death out-of-the-body state are vividly conscious of meeting dead relatives and friends, looking much the same as they did on earth, though usually younger and more radiant. Others have an awareness of such persons, but only as disembodied spirits with whom they can establish mental communication. Unknown spiritual entities, sometimes named as angels, may also be met in either of these ways. Often the visitor is told that his time is not yet; he must return and complete his spell of life in the physical body; he must get on with unfinished business, looking after a family and so on. The prospect is often met with momentary resentment since life on the other side seems so much more alluring. But on reflection he usually agrees willingly to take up again the burden of earth life. The fact that it *is* generally regarded as a burden should be a great encouragement to any who fear to die, who are clinging on to a painful worn-out physical body, perhaps beyond its due time.

The next sections, *'The Being of Light'* and *'The Review'* have already been considered in some detail.

Next we come to *'The Border or Limit'*. A few subjects reported some sort of barrier to their further progress on the other side. They felt, or were told, that if they crossed it they could not return, but they really ought to turn back and live out their physical life. The barrier was variously interpreted as water (the river Styx?), a grey mist, or a fence.

'Telling Others' about such experiences is quite a problem, because most people just don't want to hear such things; so they try to dismiss them as dreams or hallucinations or just imagination, or even evidence of mental instability. This can be hurtful to the subject to whom these events were one of the most 'real' and vivid experiences of his whole life. So he just keeps quiet, or he may anticipate these reactions and never tell anyone unless pressed to do so by a sympathetic listener. The suggestion that imagination is involved is particularly silly, for the very simple reason that such experiences are quite *unimaginable.* This is precisely what gives them their impact and their memorable quality. Nothing remotely like this had ever happened before. Even if religious teaching or reading accounts in books had prepared the individual to some degree, still the actual events were quite different in character from anything he had been able to imagine. Nor was there anything vague, irrational or dreamlike about them; certainly they were new and strange and hard to put into words, but they happened all right, and the mind and attention were, if anything, more alert than in normal waking life.

What were the *'Effects on Lives'* of these premature experiences of death? On the whole they were quietly positive effects. The subjects found their lives deepened and made more meaningful; 'they became more reflective and more concerned with ultimate philosophical issues'. Their lives tended to be more purposeful and less self-centred; they were more conscious of living in the mind and paid less attention to the body than previously; they also felt more sympathy and concern for others and a desire to find deeper meanings in life, and in general to lead better lives. Such profound and persisting changes normally result from religious conversion or the practice of yoga. It is hardly possible that they should derive from a mere hallucination, as sceptics suggest.

'New Views of Death' are a natural consequence of such experiences. The subject feels he has some direct knowledge of the event and needs no longer rely on conjecture or rather vague religious teachings. Also, and perhaps most important, all fear of death has disappeared; in due time it will be welcomed. Survival after death now seems certain. But there is no inclination to resort to suicide to speed its coming. There is a new understanding of the purpose of life on earth, as well as of death. Crude religious beliefs in rewards and punishments after death are given up for a more rational understanding.

The 15th and last element that Moody discerned in his hundreds of cases was *'Corroboration'*. This concerns checking the accuracy of events in the physical world observed from the out-of-the-body states and reported upon return. In general, they check well: doctors have been amazed at the detailed descriptions of the resuscitation attempts being made while the patient was clinically dead. Many of these accounts were given by patients with little or no medical knowledge, so they could not have been invented. In some instances Moody was able to question both the patient and the doctor, or some other person involved, to compare their accounts.

In his second book, Moody discusses four further elements related by a few subjects who had prolonged periods out-of-the-body; but none was related by only a single individual.

'The Vision of Knowledge' seems to tally with the mystical experiences reported by Whiteman, from higher levels of the after-world than those normally reached in brief out-of-the-body experiences — 'a realm of existence in which all knowledge — whether of past, present or future — seemed to co-exist in a sort of timeless state . . . a moment of enlightenment.' The subjects agreed that the experience was inexpressible and they could bring back no complete memory of it. Experienced mystics would not use the term 'knowledge' but rather 'understanding', 'insight', 'wisdom' or 'enlightenment', because this pertains to a state of supreme bliss beyond the reach of the mind. But only those who have had a glimpse of it will be able to follow this statement with full comprehension.

'Cities of Light' was a phrase used by several subjects to describe visions of celestial cities corresponding with traditional ideas of Heaven.

A few subjects encountered *'A Realm of Bewildered Spirits'*, peopled by spirits who seemed to be in an unhappy depressed state, corresponding perhaps with purgatory. They had an inward-turned hopeless demeanour and wandered about aimlessly in a dull, depressing region. Other accounts would suggest that these were suicides and other earthbound persons, dispossessed of their physical bodies before they were ready for death and the after-life.

'Supernatural Rescues' describes the experiences of a few individuals who appeared to be rescued from death after a potentially fatal accident, by some spiritual agent or being.

A short chapter provides some close parallels to these experiences from much earlier writings, namely the Bible, Plato, the Tibetan Book of the Dead, and Swedenborg. There are, in fact, many other accounts, some of which will be mentioned in later chapters.

If you want to know what life is like in some remote part of the world, say Tibet or Bali, then the best way to find out is to go there and see for yourself. If this is not possible you must talk to someone else who has been there and perhaps see his photographs. Less direct ways still are also available these days, such as travel books, films and television. Exactly the same considerations apply to gaining knowledge of this inner world. One must go there oneself, or talk to others who have been. But as yet very few know the way to get there, though many more have been taken there for brief spells without knowing exactly how it happened. Ideally one may discover the way by arduous yoga practices (See Chapter VI). Since few are prepared to do this, one must make do with the testimony of visitors and collate their stories.

I said earlier that Moody was not the first to collect case notes on people who had an out-of-the-body experience induced by a brush with death. I will mention only one of the earlier examples. In 1901 Robert Crookall published *The Study and Practice of Astral Projection,* which gave 160 brief case histories, including 21 of people who nearly died. Of all the cases, 13 reported the tunnel experience, 5 met the Being of Light, 4 went through reviews of their lives and a few others reported being with 'helpers'.

So we see that these strange happenings are reported in similar terms by all those involved. Moreover they make a very strong impression upon each individual. This inspires confidence that they are indeed reporting genuine and not imaginary events. I have not had such a near-death experience myself. Nevertheless in my brief out-of-the-body states and in other experiences, I can personally confirm a number of Moody's descriptions. These include:- the sound of bells; roaring as of the wind on exit from the body; ineffability; peace and quiet; vision of knowledge; new view of death; new understanding of purpose of life on earth; and a strong effect on my own life.

A number of people who died and came back — or who nearly died, if you prefer to put it that way — were greeted by relatives who had died previously. This seems natural and appropriate, but the determined sceptic may still dismiss it all as imagination. I do know that my own mother was there to meet my father, when he died in his sleep; but that is too personal a story to recount in detail. Perhaps it would seem more convincing if I could say that someone planned deliberately to visit some particular person who had died a few months previously and induced an out-of-the-body experience for that specific purpose. Well, I *can* say this. Robert Monroe, who was mentioned in the last chapter, did exactly this, successfully, not once but three times, visiting his own father and two close friends in the afterworld. He described these visits in his book *Journeys out of the Body.*

Dr. Richard Gordon was in his 50's when he became family physician and friend of the Monroes. Gordon had two noticeable mannerisms; he looked after himself carefully, behaving in a slow deliberate manner as if he wished to live long; then, when someone visited him in his office, he would look out from his inner office and stare intently, without saying a word (this is important in what follows).

Meeting one day, Gordon confessed he had not been feeling well, but was nevertheless going on a trip to Europe, to see countries he had not visited before. He was taken seriously ill in Europe, and, though in great pain, came home for treatment. He had terminal cancer. Accordingly Monroe felt he now ought to mention his out-of-the-body experiences. Since Gordon was too ill to be visited, it was arranged that his wife should receive a letter from Monroe to be read to the sick man when he was well enough to listen. In the event, he asked to hear the letter repeatedly before he died.

After waiting about three months for him to get settled down in the afterworld, Monroe decided to try to visit him, on a Saturday afternoon. He went out of the body yelling mentally, I want to see Dr. Gordon! After what felt like a long journey he was guided to what appeared to be a large institution. He was led just inside the door of a room and told 'Wait, and the doctor will see you in a minute.' There were several men in the room, listening to a young man of about 22 who was talking excitedly to them, but there was no sign of Dr. Gordon. It became unbearably hot, or so it seemed to Monroe, and he felt he would have to return. But first he turned again to the group of men, wondering if he should ask them about Dr. Gordon; at that moment, the young man stopped talking, turned and stared intently at Monroe, before resuming his conversation. Monroe left, regarding the experiment as a failure.

The following Saturday he tried again, but just as he left his body, calling for Dr. Gordon, a voice said, right beside him, 'Why do you want to see him again? You saw him last Saturday.' Monroe was so surprised he returned at once to the physical body and thought about the event. Going through his notes again he realised that the young man he saw *was* Dr. Gordon, looking 22 instead of 70, in the afterworld; the intense look he gave was exactly his old mannerism. Monroe had never known him at that age, but he later saw a photo of him as a young man which tallied with his out-of-the-body-vision.

Agnew Bahnson was another good friend, who happened to be an amateur pilot; he had a research interest in antigravity, and often talked of this with Monroe. The latter was on a business trip to New York and, having time to spare, lay down for a nap in his hotel room. Just on the verge of sleep, he heard Bahnson's voice saying 'There is a way to prove antigravity. All you have to do is to demonstrate it yourself, and you have been trained to do it.' Monroe knew what he meant, but lacked the courage to try it; he noted the time as 3.15. On his return home his wife told him that Bahnson had been killed trying to land his plane in a field. He asked her, 'Was it two days ago, at 3.15?' and she confirmed that it was.

Some months later Monroe tried to reach Bahnson on an out-of-the-body trip. He went out, calling for Bahnson, and travelled in darkness until he was stopped in a small room, again rather dark. Shortly Bahnson appeared and recognised him, saying, 'Bob, you will never believe all the things that have happened since I have been here.' That was all, but it was just the kind of thing Bahnson would have said, with his intense interest in new experiences.

Later, Monroe's father died, after several months of paralysis and inability to speak, following a stroke; he was 82. It was a few months before Monroe attempted a visit, concentrating on the personality of his father to direct him. Again he travelled a long time in darkness, then stopped in what looked like a hospital. In a tiny room there he found his father, looking about 50 and still rather tired. The dead man turned and saw his son, and spoke to him. 'What are *you* doing here!' Monroe was too excited to speak, and his father grabbed him under the armpits and lifted him high over his head, just as when Monroe junior was a boy. Asked how he was feeling, the father replied, 'Much better now; the pain is gone.' But with this recollection the energy seemed to drain out of him, and Monroe realised it was time to go.

I have said that in general ministers of religion have only vague platitudes to offer concerning events after death. Let me make amends by ending this chapter with an honourable exception. If you happen to enjoy classical music, you probably know 'The Dream of Gerontius', that beautiful musical setting by Elgar of a poem by Cardinal Newman. Here is the beginning of the dialogue between the Soul of Gerontius, newly dead, and his Guardian Angel.

Soul of Gerontius:

I went to sleep; and now I am refreshed.
A strange refreshment: for I feel in me
An inexpressive lightness, and a sense
Of freedom, as I were at length myself,
And ne'er had been before. How still it is!
I hear no more the busy beat of time,
No, nor my fluttering breath, nor struggling pulse;
Nor does one moment differ from the next.

This silence pours a solitariness
Into the very essence of my soul;
And the deep rest, so soothing and so sweet,
Hath something too of sternness and of pain.

Another marvel; someone has me fast
Within his ample palm;
A uniform and gentle pressure tells me I am not
Self-moving, but borne forward on my way.
And hark! I hear a singing; yet in sooth
I cannot of that music rightly say
Whether I hear, or touch, or taste the tones.
Oh, what a heart-subduing melody!

CHAPTER III

THE TESTIMONY OF SPIRITUALISM

Some readers may wonder how I have come so far with this book without mentioning spiritualism. Others may be glad that I have relied upon the more direct evidence of living people to demonstrate realms of existence beyond our normal waking consciousness, and survival after physical death. I cannot accept all the claims of spiritualism at their face value; also I do not really approve of the methods used. In my view they are harmful to the sitter, to the medium, and to the 'dead' person who is brought back to earth conditions — when this actually happens, which is less often than mediums claim and sitters believe.

In the 19th century the spiritualistic movement had a scientific flavour and attracted some eminent scientists to its ranks. It tried to demonstrate to a smug materialistic society in the Western world, the existence of another world beyond that of our five senses and the reality of survival. Many spectacular seances were held at which, beyond reasonable doubt, discarnate entities were brought back into communication with the sitters. The most usual mode of communication was by 'table-tapping', with an agreed code of raps to convey messages. But sometimes voices were heard coming from trumpets and often claimed by sitters to be recognisable as the voices of dead relatives. At other seances materialisations of spirit forms took place, usually as filmy ghost-like entities, but sometimes as substantial materialised bodies weighing up to 60 lb or so. There is even a record of a materialised spirit which picked up and carried the shrunken body of a medium. It was proved conclusively by a Belfast engineer named Crawford and by others that any weight gained by materialised spirit forms, or by ectoplasmic rods used to lift tables and to move other objects, was equal to the weight lost by the medium except for small losses suffered by the sitters. Naturally the sceptics attributed all these phenomena to fraud, but qualified scientists who made proper unbiased observations were completely convinced that phenomena they observed were genuine, however puzzling. That is not to say that every medium was genuine; some did resort to fraud, especially those who had produced genuine results but whose powers were failing under the tremendous physical strain of mediumship.

People who took an interest in these activities came to recognise the existence of a normally unseen world, and spiritualism did help in some degree to counter the deeply materialistic thinking of the time. To that extent, it had some value to offset its harmful effects. But as time went by the nature of the spiritualist movement changed. It lost the attitude of scientific investigation and took on a kind of religiosity. Innumerable spiritualist churches grew up, holding rather sentimental and amateurish services. Hymns were sung and the medium gave an address alleged to be from a spirit guide but usually quite vague and banal in its spiritual message. Then followed a period of 'readings' intended to comfort bereaved worshippers. The proceedings were supposed to be organised by a spirit guide, who would bring a number of recently deceased 'spirits' to communicate through the medium in turn. Sometimes a 'spirit' would 'control' the medium, entering her body to some degree and speaking from her mouth in his or her own voice, which would usually be recognised by some member of the congregation. More often, however, the medium retained control and would 'see' the 'spirit' clairvoyantly and would describe it; then she would speak in her own voice a message from the spirit to the living relative present, often identified by a Christian name only. This individual would be invited to respond and to say if the message was understood. Occasionally proof of identity was offered; the 'spirit' would mention some uncommon incident in its recent life, which might be remembered by the relative, or could be verified from family records. All this evidence would be accepted at face value by the credulous congregation.

In fact, alternative explanations are possible. The simplest is outright deliberate fraud, sometimes by a formerly genuine medium. Frequently the descriptions were so vague and the messages so general and trivial that someone in the congregation, desperate for a message from a loved one, would accept the inventions of the medium as genuine. Some people might even claim that if the individual was comforted and reassured then the fraud was harmless. Much more often, as I believe, fraud was perpetrated by the medium but quite unconsciously and innocently. An amateur clairvoyant with no proper training and little experience would receive an impression from outside herself, but would be unable to determine its source. So that source could easily be the bereaved relative present making a strong mental picture of the deceased person and imagining the sort of message that would be given. A sensitive medium could pick up these images telepathically, but would in all honesty attribute them to someone from the other side. Even an 'evidential' message, concerning some event forgotten by the recipient, can be explained in this way. The memory of the event might have passed beyond recall but would still be present in the unconscious mind whence it could be plucked by a sensitive medium, just as it could be recovered by a competent hypnotist.

There are other possibilites, but to understand them we need to use information that will be given in a later chapter. There are non-human entities living in the after-death world and for some of them it is an exciting experience to make some contact with the physical plane which is normally as inaccessible to them as their world is to us. Such an entity might therefore impersonate the deceased person in order to make contact with a medium and perhaps even take over her body briefly. Such an entity might even deceive a spirit guide who was quite genuinely guiding the proceedings. Another possibility is even more gruesome. As we shall see, the after-death life proceeds by its proper stages and in due course a second 'death' occurs when the individual passes on to a more spiritual stage, on a 'higher plane'; this phenomenon is indeed generally recognised by spiritualists. However, what few of them realise is that the discarded astral corpse or 'shade' is not inert like the dead physical body; it retains a dim life and intelligence of its own for a while before it disintegrates. During this period it may acquire a little spurious vitality at the expense of the medium whose body it invades. It is clearly unwise for a medium to make her body available to entities of these classes and harmful if she does so; but having embarked on this strange career she may have little choice in the matter.

Finally, the phenomenon may be precisely what it purports to be, a genuine return to a sort of borrowed earth life by a recently deceased person; he or she may indeed be impelled to return by the uncontrolled sorrowing of the bereaved. In this instance, the phenomenon is genuine and would be scientifically evidential — if only the other possibilities could be ruled out. But, it then has to be said, it is unhealthy and should not be encouraged. Death should be a one-way journey; the person who dies a natural death of old age or fatal illness should be finished with earth life and be ready for a new one on the other side of death. Returning, whether self-willed or in response to a powerful call from a living relative, is unnatural and a hindrance to proper progress; it may be deeply disturbing and harmful to the individual involved. Likewise, exaggerated mourning and a selfish craving for physical reunion with the departed is also unhealthy. Some sadness is natural but death is universal and has to be accepted. Love is best expressed by prayers for the safe and unhindered progress of the departed in the new life. Reunion *is* possible and permissible on that level, though not at the earth level, and frequently occurs in sleep though it may not be remembered. For the rest, loving memories should suffice.

I shall now retell briefly in my own words a modern story that illustrates the best side of spiritualism, the rescue circle; it also brings in other interesting facets of the after-death state, earth-bound souls, apparitions seen by several people, ghosts that talk, hauntings, premonitions and the phenomenon of rapport. I shall also add my own interpretations of events. The story has been told in great detail in the Corgi paperback, *The Ghost of Flight 401,* by John G Fuller. It started with a flight of an Eastern Airlines L-1011 3-engined jumbo jet on the night of 29th December 1972. The story also illustrates how difficult it is to foresee and avoid accidents. They are seldom due to a single foreseeable danger, but to several unforseen minor things going wrong at the same time. The plane in question was flying from New York and was preparing for its landing approach to Miami. The landing gear was lowered, but there was some uncertainty as to whether the nose gear had descended properly. The operation sounded normal but the green indicator light did not come on. The nose gear was raised and lowered again, and the crew were practically certain that it had locked correctly and that the fault lay only with the indicator light. Captain Loft radioed Miami control for permission to climb slightly and circle while they checked and this was granted. The indicator light was in an awkward position below and between the pilot and co-pilot; each of them in turn tried to release the light fitting to change the bulb, but it had jammed and they could not get it out. To free both hands the Captain put the plane on autopilot and bent down to wrestle with the light fixture. As they were still unable to open it, Second Officer Repo was sent down into the 'hell hole' to check visually that the gear was locked in position.

Meanwhile, one of them must have bumped the control column without noticing it, and this disengaged the autopilot and left the plane flying itself with no controls; in this condition, unnoticed by the preoccupied crew, the plane steadily lost altitude. A brief warning chime should have sounded and doubtless did but, as later tests showed, it was too faint to be heard by crew wearing earphones; no associated warning light was provided. The true altitude would have been shown on the pilot's instrument panel but he was too busy to check it. This was in any event the duty of the First Officer and co-pilot Stockstill. Now another fault intervened; unknown to any of them there was a slight mis-match between the duplicate controls of pilot and co-pilot, such that the co-pilot's instruments did not register the accidental disengagement of the autopilot on the Captain's side and they still registered the 2000 ft altitude set for the autopilot. The co-pilot could not see the other set of instruments from his seat. The plane's loss of height was noted on radar by the airport controller, but he was too busy directing other planes to take immediate action. When he had time to look again, it was too late; the plane's blip had disappeared from the radar screen and the plane had crashed into the boggy Florida Everglades. It was an inky-black night so the plane's descent could not be seen from the windows; it seems that neither passengers nor crew had the slightest warning that disaster was imminent. The plane broke up on impact and wreckage was spread over a distance of 1600 ft. Yet miraculously 77 people out of the 176 on board survived, though most of them were injured. This was the first crash of a jumbo jet and it need not have happened. The nose-gear when recovered was found to be securely locked in postition, as the crew supposed, so the plane could have been landed safely. Rescue operations were extremely difficult in the dark and the wet boggy terrain but proceeded relatively smoothly nevertheless; the book describes it all in harrowing detail. Of the flight crew, 1st Officer Stockstill was killed on impact; Captain Lofts told a rescuer that he was dying, and died before he could be got out. 2nd Officer Repo was rescued but died of his injuries 30 hours later without being able to speak coherently.

The author of the book later interviewed many passengers and airline staff — anyone who could throw any light on the disaster and the supernatural events that followed. Thus some interesting premonitions came to light. For example, on an earlier flight a stewardess had such a vivid premonition of a crash of one of the L-1011 fleet, between Christmas and New Year, that it made her quite ill. Three other stewardesses testified that she had told them about it at the time. She also foresaw that she would narrowly miss being on this flight and, although she had forgotten it by the fateful date, this also came true; the cabin crew was rescheduled at the last moment.

Several passengers reported that they felt they had died in the crash but came back to life again in their injured bodies. The author gives no details, but refers sympathetically to the work in this field of Dr. W. Moody and Dr. E. Kubler-Ross quoted already in Chapter II, and to Dr. Ian Stevenson's reincarnation studies to be mentioned in Chapter VIII.

Before relating the subsequent events, let us stop to consider the feelings of the three flight crew who died. They were experienced dedicated officers who were devoted to their job and to the nearly new aircraft they flew. Now they had crashed it and bore some responsibility for nearly 100 deaths. They must have felt some guilt for, after all, in those critical minutes thay had not been paying proper attention to their paramount duty, flying the plane. Probably they were confused and anxious to make amends in any way possible. So they haunted the planes and did in fact give warnings of potential disasters ahead. A dramatic example found its way into the newsletter of the Flight Safety Foundation, a publication sponsored by aviation insurance companies. On Plane 318 of the L-1011 fleet a stewardess working in the galley down in the belly of the plane glanced into one of the row of ovens and thought she saw a face looking out at her. Scared, she fetched another stewardess to check. She also saw the face which was that of Repo. Then the Flight Engineer of this Plane 318 was asked to investigate also; not only did he see the apparition bit it spoke to him, saying, 'Watch out for fire on this airplane'. Later this plane was in Mexico City when a fault developed in one of the three engines. The crew was given permission to fly it out on two engines for an engine change. But at take-off a fire developed in one of the two remaining engines. With almost unbelievable skill the crew managed to turn the plane and make a safe landing on one engine, despite the thin air of this high altitude airport. The newsletter goes so far as to hint that the ghost aircrew may have helped.

The author of the book had great difficulty in collecting other reliable first-hand reports of incidents. Eastern Airline was totally uncooperative, trying by every means to suppress and play down reports so that employees were scared to talk. The author was fortunate to find several girls who were interested in psychic events and willing to seek out key people and arrange for him to meet them. He also found a couple of airline personnel who were themselves mediums and serious members of a spiritualist rescue circle. It is clear that this experienced writer only included reports which he felt certain were authentic.

Many of the hauntings took place in the service galley. On one flight two girls independently were aware of a strange presence while working alone in the galley. Another felt an intense chill despite the heat coming from the ovens. A subjective sense of cold often accompanies psychic happenings. More dramatically, another girl watched in fear and fascination as a face, complete with hair and glasses, slowly materialised out of a puff of vapour. Loft and Repo each appeared in a First Class seat while boarding was proceeding. These apparitions looked just as solid as living people and were seen by flight attendants, cockpit crew and passengers, and in one instance by a Vice-President of the Airline. These events occurred for about 16 months after the crash, then Loft appeared less frequently, and about April 1974 the visitations ceased. Stockstill never appeared at all. Loft never spoke, but Repo gained a reputation as a friendly, good ghost who often spoke to give warnings or helpful advice.

Most of the events took place on Plane 318, but some on two sister planes of the fleet, which had also been fitted with undamaged non-structural equipment like ovens, elevators and radios salvaged from the wrecked plane. These evidently provided 'rapport' for the ghosts.

Later, this salvaged equipment was removed from the planes. It is clear that however much Eastern Airlines refused to admit the incidents publicly, senior officials did take them seriously. They understood the rapport phenomenon well enough to sacrifice valuable equipment in an attempt to break the links and end the hauntings.

To make themselves visible, people who have passed over have to materialise, to draw physical matter into their subtle bodies; this can be done to varying degrees and in different ways. The easiest way is to take it from a willing medium. When none is available some other unsuspecting person may be used, with unrecognised ability to release matter for a materialisation. Experienced entities, or those who have enlisted the help of guides, can learn to draw physical matter harmlessly from inanimate objects. The materialisation may be so slight that only the empty shell of the figure is built up. This kind of partial materialisation merges by stages into the full materialisation that Loft and Repo managed to achieve, into a solid opaque figure indistinguishable from an ordinary person, solid to touch as well as to sight. What certainly is remarkable is that the recently-dead Loft and Repo, who had no previous knowledge of psychic matters, could in fact achieve such perfect materialisation. It was presumably due to their very intense desire to make themselves known and useful which perhaps attracted help from the other side.

Four of Fuller's new friends told him they proposed to hold a 'soul rescue' session and invited him to be present; the meeting was arranged for March 18th, 1974, some 16 months after the crash. Since Captain Loft had already ceased to manifest it was only necessary to concentrate on releasing Don Repo from his earth-bound state. They entered a reverent state of meditation, trying to raise their consciousness to a higher level, and invited Repo to make contact. Shortly two members of the group were aware of his presence and one of them, a woman, became intensely involved, seeming to live through the crash as Repo had experienced it, including severe head pains corresponding to his injuries. They explained to him that he was dead and he ought to dissociate himself from earthly affairs and go on into the new realms where he now belonged. They told him that many helpers were available to him, that he should look towards the bright light he would see and make his way towards it with their help. The two who had made contact sensed that Repo understood but was reluctant to do as they asked; nevertheless he did eventually turn away towards the light, and the session came to an end — leaving the active participants exhausted by their efforts.

In mid-July 1974 a flight officer boarded his plane, No 318, early and found two terrified stewardesses on board who had just seen a ghost. The officer comforted and reassured them explaining that he knew all about these incidents and had for some time been considering attempting to exorcise the plane from these visitations. He was a deeply religious man who sincerely took the Bible as his guide though he had made some study of psychic phenomena. He felt that exorcising a plane should be feasible, that he was comptetent to undertake it, and that it had better be done discreetly and secretly; the airline was unlikely to accede to a formal request. So he decided to go ahead then and there; it should be mentioned that he had no knowledge of the earlier soul rescue attempt. He went down alone to the galley, taking a cup of water. As he arrived there all the lights on the plane started flickering as if to signal that he was expected. So he started his own informal version of an exorcism service. Naming the water as symbolising the Blood of Christ, he sprinkled it about the galley by way of sanctification. A strong wind blew up in the closed galley and the air became intensely cold. Repo materialised briefly and was told firmly that he was dead and had no business there. 'In the name of Jesus, by His Blood, you are begone from here. I am calling an angel of light, and he shall take you to the place that you belong. This is scriptural.' Then Repo disappeared, an intensely bright light appeared where he had been, and the exorcism was over.

So three quite separate efforts had been made to end the visitations; which was the successful one? I think it must be assumed that hauntings would have ended sooner or later in the natural course of events, just as the appearances of Captain Loft had already ceased. But all three efforts probably helped towards termination of the events. The removal of rapport-making equipment took place over a period and no dates are recorded. The soul rescue was in March 1974 and appeared to be successful; but it would not be surprising if the anguished soul reverted to his strong habit and returned a few more times. The exorcism probably completed his rescue, since it is the last recorded event on the planes, although Repo made contact again in other ways.

Fuller started as a complete sceptic, but he was greatly impressed by level-headed down-to-earth engineers and others, who nevertheless could speak in an equally matter-of-fact way about psychic phenomena, and who themselves were mediumistic. What finally persuaded him that his book was worth writing were the final events of this saga. Some of these friends he had made during the investigation persuaded him to attempt to contact Repo using the ouija board. This was successful and Repo conveyed some clues that were meaningless except to his wife and daughters.

CHAPTER IV

DAILY LIFE — HERE AND THERE

Chapters I and II presented evidence that life *is* possible in a non-material condition while the body is asleep or in other ways inactive. If this is possible when the body is still alive, it is only a small step to concede that it is also possible when the body is finally lost, at death. Not only is it *possible*, it is very *probable*.

If we have this non-material body while we are still alive, if it can operate on its own while linked only by a thread to the material body, why should it not do so when this thread is severed? Chapter III presented evidence that life *does* continue after death. Thinking along these lines makes nonsense of the idea that we are completely 'snuffed out' as individuals when we die. This notion put about by diehard materialists seems silly when we think of all the evidence for survival. Moreover, we have mentioned only a tiny selection of what evidence is available. So from here on we shall take as established that life *does* continue after death.

It has often been said that death is like sleep. Every night we are brave enough to let go, to allow ourselves to slip into unconsciousness, because we are confident that we shall safely wake up again some hours later. One of these days we must be brave enough to let go into that longer 'sleep' from which we shall 'wake' into the 'World of Imagination' instead of on earth; that is the only difference. I shall return to this theme in the final chapters.

I shall call the body we use in the first period after death, our 'Body of Light'. In many books it is called the astral body and the world it lives in is called the astral plane. I shall not use these terms here because I do not wish to arouse preconceived ideas and because I believe these ideas are to some extent wrong. Perhaps I should say that these terms are all too easy to misunderstand. The astral plane, or astral level of consciousness, is often thought to be concerned exclusively with feelings, emotions and desires. But in astral experience, emotion and thought go together unavoidably. In any event, there is no doubt at all that we still have the use of our minds after death. Indeed, thoughts are inclined to be even more vivid and free-flowing because they are no longer channelled through the brain which can tire. This is a world of thoughts as well as of feelings and emotions. I call it the 'World of Imagination' because its filmy substance is so obedient and responsive. It is 'ideoplastic', moulded by ideas; just to think is to create. The imagination can run riot, can ask for and have anything it wants.

Would it not be a good idea, then, to prepare ourselves for the great adventure which must inevitably come for every one of us, when the time is ripe. Otherwise it will be as if we were transported to a strange land, not knowing its language and customs. How confused and bothered we should be, not knowing how to behave, how best to spend our time. When we *do* visit a country that is new to us we take a guide-book or we travel with a guide who can explain everything. Let this book prepare you in advance for 'Our Last Adventure'.

First we now have to drop this analogy of going abroad. It is not so much a new *place* we are 'visiting'; it is more like taking up a completely new job and having to adapt ourselves to it. We shall enter upon a totally new way of life in our Body of Light and it will take some getting used to at first. In this and the following chapter I shall try to tell you what I think it will be like.

But how do I know? You are entitled to ask and I will try to explain. First of all we can learn a great deal by just thinking about it. We recall what we already know and apply our imagination to work out the details. This is a kind of applied common-sense; perhaps 'creative meditation' is a better description of the process. Any of you who have practised meditation should be able to understand this. One method of meditation is just thinking around a subject using a disciplined mind that does not wander aimlessly.

By reviewing what we know already, dwelling upon it and then pressing beyond, our information orders itself better and we gain fresh understanding. In other words, this process of creative meditation opens the mind to a higher wisdom; it encourages intuition to flood down into the mind to enrich it.

There are people who have been working along these lines every day for decades, slowly developing their mental abilities and intuitive insights beyond anything we can readily imagine. Some of them, by using advanced yoga practices, have trained themselves to leave the physical body at will and use their Body of Light just as we use our material bodies, so they can visit the inner worlds that we inhabit after death and remember what they have experienced. Other people are born psychics and they too can 'see' these inner worlds, with varying degress of proficiency. Some of them have written down and published records of their experiences. Their accounts differ in detail but agree in broad outline so we can rely on these books. (See Chapter VI)

We must remember the many countries where the predominant religions are, for example, Hinduism, Buddhism or Mohammedanism, where religion plays a larger part in daily life than it does in the West. Among these people belief in life after death is taken for granted because it is clearly taught by their religions. It would be surprising if the consistent beliefs of thousands of millions of people were totally untrue, as some materialists would have us believe. The investigations of self-trained clairvoyants and natural psychics confirm what is taught in poetical language in these Eastern religions; but they provide much more detail. These people write in a straightforward, modern, descriptive fashion, though it must be admitted that there are no words readily available in which to describe these unusual experiences. We use words for *common, shared* experiences, then everyone knows at once what we are talking about. But for an *un*common happening, for some new event, we have no words ready to hand. We feel lost, perhaps even a little scared, and we can only say, haltingly 'It was something like this but not *quite* like, it was different somehow.' So we have to fall back on analogies and this is a way of talking and writing that poets use more freely.

Until a few centuries ago the people of Catholic Europe were devout and religious. They were taught, and they believed, that good deeds on earth were duly rewarded in heaven; that after death there was an intermediate state of purgatory, in which a man had to purge himself of passions and sin before he was in a fit state to enjoy the delights of heaven. So they prayed for their dead, they had masses said by the priests, to help their loved ones through this uncomfortable stage, to turn their faces towards God. But more recently, Western man has been busy securing material prosperity, developing scientific discoveries and inventions to provide ever more machines and luxuries for comfortable living. He has had no time or inclination for other-worldly thoughts; he has focused almost all his attention on things of this world, on material possessions and has neglected the inner life of the Soul and the Spirit. Knowing full well that he must die some time, he has nevertheless thrust the thought aside, until it has become almost indecent to mention death. Protestant Churches speak of it reluctantly, as if with embarrassment, and what little they say is not very helpful. This is a very sad state of affairs. People are killed in accidents, or they die suddenly of a heart attack, utterly unprepared for the new life into which they are so abruptly thrust. They are bewildered — 'lost souls' — and it may be a long time before they even realise they *are* dead and come to terms with their new conditions. This is not just unfortunate, it is shamefully unnecessary. It is not difficult to learn something about the life after death. With a little quiet thinking and some guidance we can easily imagine for ourselves what it will be like.

If you happen to have lived a very ordinary humdrum life, then it could be that dying will be the most exciting thing that has ever happened to you. If, on the contrary, you have lived an eager full life, devoted to creative activities and crowned with awards and honours, then at the very least, death will be another intensely interesting experience. It is nothing to worry about, not an event that need evoke horror or dread, nor something to push out of your mind.

So this is where we start. It seems a good idea to take a typical day in our lives and think out how it feels now, with a body, and how it might feel then, without one. This is not something to be solemn about; on the contrary, you can make a game of it and begin to enjoy working out how things will be. After all, it is all going to happen some day; there is nothing freakish or unusual about it; it just hasn't happened to you — yet. When you go on holiday, you look forward to it; you imagine happily what you will do, the many ways you will enjoy this period of relaxation. You can do the same for your last adventure; look forward to that also with joyous anticipation, instead of with horror and dread. Believe me, it *will* be enjoyable on the whole when it comes.

How do you wake up in the morning? Perhaps you need to be woken by an alarm clock to be in time for work; perhaps you are fortunate in being able to sleep until your body is fully rested. Perhaps you have time to lie awhile and plan your day's work or just daydream pleasantly. Now this *is* something you can go on doing over there; in fact you will have all the time in the world for it. What is more, you will find your imagination much more vivid, because you no longer have a brain through which thoughts have to be channelled, but you still have a mind, and it can roam free. From necessity or habit, here you must leave your comfortable bed and get up. You must lift your 100 lb or so of flesh and bone from its supine position and set it upright on its feet. An unbiased visitor from another world might well find it easier to imagine how we get about in our filmy Body of Light than to understand how we manage our heavy hunk of a body. Try lifting someone else to get the feel of the weight. Yet we do cope with it all day long, without even having to think about it. So beautifully constructed are our limbs, our joints and muscles, so well coordinated are they by nerves and the automatic part of the brain, that we can stand and walk almost without effort. But with advancing years it does become a little more difficult; joints get stiff and creak, sometimes painfully; when we walk uphill, we find ourselves breathing heavily and the heart thumping. When we fall ill, we become far more conscious of the painful body and sometimes wish we could be rid of it. Well, so we shall be, in due time, and what a joy that will be. Just imagine feeling light as air, free from all that burdensome weight, free from pain and fatigue. That alone is something to look forward to, not something to be afraid of. It will seem strange at first, but you will soon get used to it. After all, the sensation is not all that strange. We all dream and most of us remember something of our dreams from time to time, if not every night. In our dreams too, we seem free and weightless; this is especially noticeable in dreams of flying, which are not uncommon. So in your Body of Light you will feel what a joy it is to be alive. But wait a minute — all your relatives and friends are mourning because you have died. Of course it is natural that they should be grieved at losing you. But from your new vantage point you may well wonder at the situation, general though it is. You may well fell that *they* do not need to be sorry for *you*. You are the lucky one; they must all endure the burdens of earthly life for a few more years before they too are released.

On earth again, having got up, you have to start the daily routine. If you are a man, you probably shave. What a tiresome chore, every single morning! Well, that is something you will be free of, unless you want to imagine your beard is still growing. Then you wash, but in the other world there is no external dirt to wash off, so that too is a habit you can forget. It is true that, figuratively speaking, you probably do still have some internal 'dirt' that has to be cleaned away but we will talk about that later on. At some stage in the daily routine, you get dressed, and you must choose what you will wear today. In your Body of Light this is no problem, you will probably still imagine yourself clothed 'for decency' but imagination is all you need. If you want to change into something different, just imagine you have done so — and lo! there you are in fresh clothing!

Next comes breakfast. Habits are strong and are not automatically cast off with the dead material body. So for a while you may imagine yourself taking breakfast and other meals, but there is no need. You no longer have a body that gets hungry periodically and needs nourishment to have the energy for work. The Body of Light has abundant energy, which it just draws in from the surroundings, without effort. But perhaps you still *like* eating, you enjoy your food and look forward eagerly to meal times. If so, you have met one of the snags, one of the little *dis*advantages of life after death. You will get little or no satisfaction from your imaginary breakfast, so you will soon give up eating altogether. To forgo the love of food is one of the many little things you will have to re-learn in order to adapt happily to your new life. After all, when you go on holiday you probably eat strange food, especially if you go abroad. Other things are different too; you do not sleep in your familiar bed. But that's the whole idea — you go away 'for a change', to enjoy the sun and sea or mountains, or whatever you expect from a holiday. These

Z1-Q-011-03868

76903655-9

benefits more than compensate for most inconveniences. So when you make this more drastic change in your life, you must expect many things to be different and some of them inconvenient or unpleasant. But again, the compensations, the joys of this new life are that much greater.

Returning to daily life on earth, you are up and breakfasted. Some time, you probably go out, hurrying to work, going shopping, taking the dog for a walk, or just getting some fresh air and exercise. You walk some part of the way (or all of it) and you may use public transport or a car. So what is travelling like in the world of imagination in which you will find yourself? Well, you can still walk if you wish to, and the habit is so strong you probably will. Public transport and cars are not available because they are quite unnecessary, nor indeed is walking necessary. There are two other ways to travel, both accomplished just by imagination. If you want to go at a moderate rate and enjoy the scenery, then you fly, not in an aeroplane, just in your Body of Light, or on a magic carpet if you prefer to imagine some vehicle before you get used to the idea of flying and soaring like a bird. Soon you will arrive where you want to be and you just stop. Imagine, if you like, that you have a little weight and float down. There is a second swifter way to go. You want to see a friend perhaps for a chat and you do not know where to find him (or her) in this strange land, so just think of the person, and — hey presto — you are almost instantly together. Indeed, if you start off walking or flying, you must be careful *not* to picture your destination too clearly, or this second mode of travel will take over and you will find yourself already there. This experience is not so strange as it sounds; most people know it in dreams, especially as the dream ends and they return swiftly to the body, sometimes as if falling from a height. Incidentally, if you do have falling dreams and they scare you, try to remember there is always a safe landing, you are coming back home.

Returning to the imagined day in your life, you were about to leave your home. If you were just taking a walk, then that is something you can do any time on the other side. If you were off to work, you will just have to think of yourself as retired already, for there is no need to work over there. So, if your work was of a boring routine nature, which you did to earn a living, then you will be glad to be relieved of it. There is no longer any necessity to work in order to earn money, to support a home and family. There is no need for a house to shelter from cold and rain or excessive heat. Such conditions do not exist in this other world, nor do we have senses to appreciate them, even if they did exist. The Body of Light does have senses and perceptions, but their nature is different from those of the material body, as described already. Since there is no body to nourish with food, there is no need for money to buy it. But if you loved your work, if it was creative in some way, then although you do not *need* to continue it, you can do so if you like. You can just create whatever tools and equipment you require, by imagination, and carry on writing, painting, playing or composing music, or whatever it may have been, to your heart's content. Later you may discover work that is more appropriate and constructive in the new conditions, but more of that later. (See also Chapter IX)

You may, of course, be a housewife with a more homely routine. Like your husband, when you die you can retire from most of these activities, like cleaning, washing, shopping, and perhaps looking after the family. At first, you may feel at a loss without all these chores to occupy your day. Instead, you can enjoy the long rest and relaxation from duties, that you must often have longed for. Though you will be cut off from normal relationships of everyday life, with children, family and friends, you will not lose touch with them altogether; under different conditions, less intimate in some ways, but more so in others, you can still watch over them, and send them loving thoughts.

Some retirement activities and some hobbies can be continued happily, though in modified ways, in the after-life. If you enjoy gardening for example, you will be able to admire and enjoy gardens created by your friends who have passed on, and by others you did not know on earth. If you wish, you can create your own garden. But there will be no backaches from the digging that does not have to be done, or from weeding, because you need not have any weeds; and the flowers can bloom at their loveliest all the year long, even ones that on earth are killed by frost. Everything is beautiful and effortless, except for the use of clear-cut imagination; but also no challenge, as earthly gardens are, so you might get tired of it after a while.

You can pass over lunch and evening meals and think of evening activities. If you enjoy reading, you can still do this it seems, on the other side. Music to suit all tastes is available too. But what about television, that universal entertainer? I feel sure you could imagine your set, and fill the screen with your own pictures, but I doubt if it would behave as on earth, providing the scheduled programmes automatically. Possibly you invite friends or relatives round for a drink and a chat, or they invite you. The drink can be imagined on the other side, but I fear it will not prove satisfying. Certainly you can meet friends over there, but there are restrictions on whom you meet, and when, just as there are on earth, though the restrictions are different. For example, here you do not expect to meet people who have died, except rarely and unsatisfactorily, through a spiritualist medium. But there, the recently dead are the most accessible companions. On earth you cannot meet friends who live a long way distant, even overseas. Over there, distance is no barrier, for travel can be almost instantaneous simply in answer to your wish. In life you do not call on your friends in the middle of the night, even by telephone, except in dire emergency. After death the situation is reversed. You may *see* them during the day when they are awake and about their business, but they will not see you nor be aware of you except very rarely, so they cannot respond. But when they are asleep and conscious in their own Body of Light, that is the time when you can commune with them. You may have intelligent conversation together, you may give them messages or warnings, and of course, you will remember all this, but will they? The answer, unhappily, is probably not. They are using their minds, but their brains are asleep with the rest of the body and it is quite difficult to impress memories on the sleeping brain. It can be done with practice, possibly catching the brain in a receptive state at the moment of waking, but few people can do this. So if they recall anything at all in the morning it is likely to be no more than a sense of having been with you and perhaps some garbled message that may not mean anything to them. This is most frustrating for you, especially if you desperately want to pass back some important information, like where you have put your will, for instance. The only thing is to keep trying to contact other people who may be more receptive and who can relay the message to your loved one.

But do not despair if you are concerned for your children. Just keep on loving them, and meeting them at night; be sure your love will enfold and protect them even if they do not seem to be aware of it in their waking lives.

There is another restriction when you are newly dead. You may not be able to meet relatives and friends who passed on some years or decades ago. *They* can probably contact *you* and give you their love and encouragement, but they probably live at higher levels of the World of Imagination which you cannot yet reach, or they may live beyond this world (see later). Be assured that you too will move on, and that you will rejoin them eventually.

But perhaps in this earth life you spend the evening at home on your own; perhaps you want to plan your holiday. Would you stay somewhere in this country, or go maybe to one of the nearer European countries, France, Belgium, Spain, or further afield to Switzerland or Italy even Greece or N. Africa? Anyway you would study brochures and arrive at some compromise between what you would like and what you could afford. Will it be possible to indulge a love of travelling, after death? The answer seems to be yes if you still want to do so. Your new world will be different, in another space from that of the familiar material world. But it is said that its lower domains are a fairly faithful replica, or counterpart, of our earth, with all its natural features and man-made cultivation, developments, buildings and so forth. As we have already noted, travelling is free and easy, so why not explore the world, without the discomforts and inconveniences of cars, coaches, trains or planes? I remember my dear father, who had not travelled much, saying that he planned to do just that, on the other side. I wonder if he did; so far as I am aware he never came back in any way to tell me. But I do know that my mother, who had passed on many years earlier, was waiting to welcome him into that world. I feel sure they would have wanted to stay together but as she was not so keen on travelling I expect they reached a happy compromise.

As a final possibility, for a typical day that we are re-living together, maybe things had gone sour for you in some way, and you were considering the radical step of leaving the country, not just for a holiday, but for good, emigrating in fact. Could you bear to uproot yourself, leave relatives and friends behind, and settle in a strange country you have never seen before? If you have never thought along these lines, then I beg you to do so, as a kind of preview of what is to come. For one of these days you are going to leave for an even more drastic change of life, in an even stranger new world.

And so to bed, as Samuel Pepys put it. After some 16-18 hours awake, most people find their body feels irresistably tired; they want to go to bed and sleep. The body needs all that time at night to recuperate, to recharge its batteries, so to speak, ready for the exertions of the day ahead. But it is only the material body that feels fatigue; the brain may feel weary too, but that is also part of the material body. The mind behind it does not tire, nor do the emotions and the spiritual nature. In sleep you remain conscious in your Body of Light, but next morning you probably remember little or nothing of the night's adventures. Nature wisely draws a curtain between these two parts of your lives. While you are living on earth you need to get on with living your waking life to the full, making the most of it while it lasts. It would not help if you were burdened with the interests of a second kind of life lived in sleep. It is not wasted; you will pick up the threads of that sleep life in due course, and remembering it will help in adjusting to the permanent loss of the body. Some very old people, especially if they have a long terminal illness, may find themselves able to lift that curtain a little and so prepare themselves more fully for what is to come.

CHAPTER V

GETTING USED TO LIFE WITHOUT A BODY

In the last chapter I emphasised the advantages of losing one's body. They are, indeed, many: freedom from aches, pains and fatigue; weightlessness; freedom from burdensome responsibilities towards the job, family, homes and possessions, and whatever else. All this can compensate for a lot of bewildering difficulties in adjusting to a totally new way of life. But it cannot be denied that many people do find themselves in deep trouble. Those who scoffed at the very idea of any existence beyond death, who seriously believed that death brings total and instant destruction of the personality, such people feel utterly lost and perplexed. Things are even worse if they are cut off suddenly by an accident, in the prime of life. Such individuals often fail to realise that they *are* dead. Having just read descriptions of how different the new life is, you may find this hard to believe; but it is true, all the same. Our habits are ingrained in the brain and in the animal consciousness of the material body, but they persist in the mind after death and the habits of a lifetime are strong and hard to break. So it is easy at first to go on doing all the habitual things, admittedly in imagination only, particularly if you have never even considered the possibility of any life after death. Such people are so 'earthbound', so wrapped up in themselves and their daily routines that they simply fail to observe, for a long time, that things have changed. They are of course puzzled and frustrated that their still-living family and associates do not seem to notice them and make no reply to conversation. Have they been 'sent to Coventry' for some obscure reason? Naturally, material objects that they try to move, in fact do not move, but in their imagination they *will* seem to move the counterparts of these objects in their new world.

If several people were killed together, as so often happens in road accidents, for example, they may find themselves still together and in the same deplorable condition, able to communicate with one another and just reinforcing each other's illusions about the state of affairs (See Chapter III). Such earthbound souls, in their self-centred perplexity, are likely to reject and turn aside from those who proffer help and try to explain the way of things. Such helpers may be recently dead people who have been through similar experiences themselves but have been shown the need to wake up to the facts and come to terms with the new lifestyle; or they may be caring, living persons who make such soul-rescue their mission and who work this way when they are asleep, or in trance or meditation.

After reading these words of mine, the ideas will remain in your mind, even if you find them hard to believe. So you are unlikely to have this distressing experience of not believing that you are dead and you will at once start trying to adapt to the new circumstances. That is one reason why I have written the book, under the direction and guidance, as I firmly believe, of Those who have gone ahead, but who have the welfare of humanity at heart. They miss no opportunity to show us where we have gone wrong, if we will only listen, and to lift us out of the mess we have made of our civilisation. So please do finish reading the book, and then pass it on to sceptical friends. It would be even better if they could be persuaded to buy the book.

Before going on, I must mention that the most unfortunate are those who die by suicide. They are condemned by their forbidden act to live out the rest of their allotted span of life in a sort of limbo, still belonging to this world but unable to make effective contact with it, because they have forcibly removed themselves to the lower regions of the next. They truly live in hell, but it is of their own making, and fortunately it is *not* eternal. Never, never be tempted to commit suicide. However terrible your earthly entanglements have become, however logical a solution suicide may seem, do not be tempted. Believe, as is true, that the resulting state will be even worse to endure than the present one.

But enough of such gloomy thoughts. Let us assume that you have died a normal peaceful death, that you recognise your condition and you are ready to adjust to it. You may have noticed that in the last chapter I mentioned a number of activities that are natural to the Body of Light. But I inferred that others you might wish to indulge in were really unnatural to your new state: they pertained to earthly life, and were continued out of habit, eating and drinking for example. Also I implied that although these actions could still be performed in imagination, they were unlikely to give much satisfaction, because you no longer had the physical senses of taste and smell which made them enjoyable on earth. They become hollow pleasures and lose their enticement. So it is just as well to recognise this state of affairs, and to wean yourself gradually from such inappropriate activities. If you were a hearty eater, somewhat dependent on alcohol for stimulation, and perhaps tobacco for release of tension, then you may find it hard to give up these things. Perhaps you derived pleasure from sexual indulgence but in the after-death state you have no body and the only way is to behave in a manner proper to your new conditions; the only way ahead is to disentangle yourself and break free. If you rebel against this, if you persist in clinging to old habits, then you will be making a hell for yourself. It has been truly said there *is no hell,* unless *we make it for ourselves.*

What purposes are served by our being born on earth? There are several, but here I want to concentrate on just one — getting to grips with things. I have been at pains to explain how much easier life is in the realms beyond death. At those levels of existence, whatever we want can be created by almost effortless imagination. We can behave, if we wish, like the spoiled children of rich parents; I expect most of us could drift along happily for centuries in that somewhat dreamlike existence. But I think there would come a time when we should welcome some challenge, something a little more difficult, some opposition against which to test ourselves. In fact we should be glad of some real work to do; that of course, is what life on earth is all about. The things of earth do not respond to our wishes; if we want something, we must make it, using physical effort as well as thought. Far from responding to imagination, earthly rocks and ores are hard and unresponsive; they have to be worked on laboriously with hands and tools, or be smelted by fire, before they become useful. Our present complex material civilisation stands as a monument to our achievements, but it has become too interested in material things.

The accumulation of goods and possessions and luxuries of all kinds, or their equivalent, money, has become an end in itself. Instead, these things should be a means to an end, namely to provide us with more leisure. We do indeed already have more leisure than we had only a few years ago, with shorter working hours and longer holidays. But many of us squander these opportunities in a round of entertainment. Their true purpose is to provide opportunities to acquire wisdom and to grow in spiritual stature.

Then in a single stroke, by death, we are stripped of all our possessions. The monks and nuns and the holy men of the East renounce them voluntarily, but we are compelled by death to discover that they are not essential. Life after death is a rehearsal for the spiritual progress that humanity *must* make in due time. This may seem a mystifying statement, but I hope to make my meaning clear in the course of this book. An almost complete reversal of our customary attitude to life is required of us in the after-life. On earth we are in the main, and quite properly, concerned with exterior things; we are outward-looking in our behaviour, though not all the time, of course, and not to the same degree with everyone. We just have to be like this to make a success of our lives. After death we must change to become largely inward-turned, introspective, concerned mainly with our own inner thoughts and feelings. The situation is really far more complex than these few sentences suggest; it is impossible to crystallise it into such a clear-cut pattern. If you like, this is a sort of shock-treatment exaggeration to jolt your thinking along new lines. Of course, on earth, we do have periods of reflection, of looking at ourselves; on the other side, we do not become totally self-centred and withdrawn. We can have no more traffic with earthly *objects,* but we still need to relate to other *people.*

To clear up another possible misconception, it is useful at this stage to reflect that your material body is, in a certain sense, an earthly object and not really 'you' at all. I know very well how strongly most people identify with their bodies when suffering from raging toothache or other forms of acute pain. But a few people can manage to banish intense pain from consciousness, especially with some help from hypnosis or acupuncture. In more normal circumstances, it is not too difficult to develop the practice of regarding your body not as yourself, but as something you *have,* to use like an intimate tool in the material world. If this seems hard to grasp at first, I must make it even harder by pointing out that the body includes the brain. This also is a tool, a kind of built-in computer, used by the non-material mind, the more enduring part of the personality which does survive death. When this comes you will be forced, instantly, to manage somehow without body or brain, so it is a helpful exercise to try to realise, in advance, that this is largely the way things are right now, though we may still lust for the joys of earthly things. Of course, you do have a body that has evolved from the animals, and its animal-like nature inevitably forces itself on you to some extent, but it should never be allowed to enslave you. On the contrary, the body should be your slave and if you work harmoniously together it can become your obedient and trusted servant. Quite definitely you are not just an advanced sort of animal, a 'hairless ape' with a superior brain, as some authors would have us believe. You are something quite different from this. You are a self-conscious spiritual being, a Spark of the Living God, with a mind of your own. If you can manage to remember this, all the time, then life will have a new meaning for you. You will acquire a sense of security with this faith and a wondrous understanding of who you really are, and where you are going.

After death all this will become a self-evident matter of experience beyond possibility of denial. You cannot avoid it, so why not believe it now, and enjoy a foretaste of this bliss while still on earth? If you persist in this, you will find your way of life changing slowly; it will become serene and more deeply enjoyable. What is more, when the time comes you will adjust swiftly and easily to the conditions of life after death because you have already accomplished much of what normally has to be done, painfully, in its early stages. I am not advocating any sort of pious, holy attitude, nor retirement to a monastery. All this can be done while living a normal life; it is just a matter of enlightened common-sense. You have to get that way sometime, so why not now?

I am well aware that this is a counsel of perfection, and that you are unlikely to be fully successful. So brace yourself for some sorrows at the start of the new life. But at least you will know the sort of situation you are likely to face, and the sort of attitude you could adopt to get through it speedily. Be assured, moreover, that it gets better all the time, if you face your fate instead of cursing it. Things are so arranged that our coarsest desires, our more animal-like ways face outwards, so to speak, so that they are dealt with and purged away first. Having worn out and cast off our baser desires, we can 'rise' to happier domains of the World of Imagination to deal with the more insidious, but less shameful, attachements that still hold us back from heavenly bliss.

But first, let us return for a moment to the materialist, who sincerely believes that at death he will be snuffed out like a candle, as an individual. Some philosophers have said, in true humility, that they are not worthy to survive, that the Universe would not be enriched by their continued existence. This is a view that deserves to be considered. Indeed I would agree that in a certain sense they are right, if they are regarding themselves as their personalities. I do believe that this lower part of ourselves does *not* survive, except as a minute part of the total 'Memory of Nature'. But unlike the materialist, I do not believe that it disappears instantly at death. Rather, it slowly dissolves away in the afterlife, as its purpose is completed. But this is not the end of us. I feel sure that the true inner self, 'YOU' does persist, eventually to reincarnate again.

CHAPTER VI

PROGRESS IN THE AFTER-WORLD

Up to now I have relied upon the testimony of amateurs in this field of research into the inner worlds and the life beyond death. I have done this deliberately, to inspire confidence, because these amateurs are ordinary people like ourselves. Under slightly different circumstances you or I might have done what they did. Perhaps they happen to be less closely attached between their physical and higher bodies, so that it is easier for them to learn the art of leaving the body. Or they may be better able than most people to impress the clear memory of an out-of-the-body experience upon the physical brain. Anyhow, induced by exceptional stress, or by drugs, they can recall such an experience, though maybe only once; others find it interesting to collect and publish accounts of such events.

A few of the authors I have quoted may not regard themselves as amateurs. Muldoon made repeated trips but he had little control over where he went and he never seemed to get beyond the physical world, or what Whiteman calls the duplicate world, into higher realms. Monroe made hundreds of trips, many of them self-induced, though it sometimes took him as long as an hour to leave the body. In most instances he went where he was taken, but on some occasions he was successful in his deliberate attempts to visit specific living or dead persons. Whiteman's experiences numbered thousands over many years of his life. He took a remarkably passive attitude towards them; he seldom if ever deliberately induced a separation and refrained from attempting to exercise any control over events beyond trying to will himself to higher levels if he became trapped in lower realms by his own fixations. I have met him and we have exchanged long letters. He is an exceptionally spiritual man though quite matter-of-fact about it; he is in fact a university professor and something of a mathematical genius. He accepted every experience as a precious gift sent to teach him something more about the spiritual life. Much as I admire him, he remains, for my present purpose, in the amateur class because he made no attempts at deliberate exploration of these inner worlds, but was content to note and record what he observed.

For further details, I shall rely upon professional explorers of these inner realms. First we may turn to ancient Egyptian civilisations. Some of their splendid temples still remain intact. The priests and priestesses who once officiated in these temples were trained to leave their bodies in sleep or trance and to remember what they saw of the inner worlds. So they were quite familiar with events after death and knew something of the higher purposes of birth and death. For an impression of life in the palaces and temples of that civilisation I commend Joan Grant's book, *Winged Pharaoh*; it is a novel, but is probably based on memories of a previous incarnation. Some part of the knowledge of these priests has come down to us in the *Egyptian Book of the Dead* in which information about the after-world is presented in poetical and symbolic language.

A remarkably similar book is the Bardo Thodol, or *Tibetan Book of the Dead*. This sets down the first-hand understanding of Tibetan lamas, doubtless trained in their monasteries in much the same way as the priests of ancient Egypt. The book was translated into English by Evans-Wentz and was first published in 1927 with several lengthy forewords and introductions by distinguished writers including Jung. The book is written in a sort of poetic cipher language, as a prolonged funeral service addressed to the newly-dead person and intended to be read over a period of many days, to guide the soul on its way. But it is also intended to be read by the living, for guidance and instruction towards leading a more profitable life and easing the passage through the early stages after death. For example, it is pointed out that in life there are many minor deaths to be dealt with, before the final one marking the end of life on earth. Therefore every moment should be faced as if it were one's last on earth. It is remarkable that exactly the same advice was given by the Mexican Indian 'sorcerer', don Juan, to his pupil from the Western world, Carlos Costaneda, who was being taught 'the warrior's way' of living. The Bardo Thodol is unique, however, in covering the whole cycle from death to rebirth. Like the ancient Egyptians, the Tibetans believe implicitly in reincarnation. It is sometimes said that they believe in transmigration of souls into animal or insect bodies but the Evans-Wentz book explains that this notion is based on a misunderstanding. Transmigration does not occur literally, but a man who has lived unwisely may be condemned to incarnate into a lowly *human* body that will display the characteristics associated with some particular animal or insect.

In India, too, first-hand experience of these matters has been available for thousands of years to the holy men, the Rishis and Yogis. For them, it was just an incidental part of a wider understanding. Yoga means union — union with the Divine, which can only become effective in its fulness in realms beyond even the heaven world. So to reach these sublime levels the yogi must first become a perfected human being; he must gradually discard all the more earthy parts of his nature, and bring what remains under the strict control and discipline of his higher self. We shall return to these ideas in the next chapters; for the present it is enough to realise that these yoga practices give the seer the freedom to explore the inner worlds, including the condition of souls after death.

During the last hundred years, people from India and the West, guided to take up yoga practices for self purification, have made this knowledge more freely available; this has enabled them to explore the inner worlds themselves, but in addition they become worthy to receive the Ancient Wisdom from Those ahead of them in evolution, often named the Masters of the Wisdom. Their teachings were at first channelled through the Theosophical Society and were published over the years in a number of books. More recently, this knowledge has become widespread and has been taken up and amplified by other authors. The early workers were men and women of a higher calibre than most of those already mentioned. They were persons of iron will and immense self-discipline, prepared to sacrifice most of what makes life enjoyable to ordinary people. Under the guidance of a guru, they undertook arduous yoga practices over many years, to subjugate the personality and to allow the higher self to take complete control of their lives. There is nothing secret about these practices; they were set down, in Sanskrit, by the seer Patanjali, some 200-400 years B.C. but it is one thing to read his *Yoga Sutras*; quite another to practice them diligently. Some of the results of these yogic exercises are psychic powers, and freedom in the inner realms of feeling, thought, wisdom and intuition. In India and other Oriental countries there were many such yogis. Some of them became gurus themselves and accepted pupils; a few recorded their experiences in their own languages.

I prefer to confine myself to writers who have described what they know in English. As scientists have discovered, the world of nature is extremely complicated. This complexity extends also to the inner worlds, as the writings of H. P. Blavatsky (founder of the Theosophical Society) suggest. Her accounts seem to me to be excessively involved and unclear. Therefore I have chosen to draw instead upon the writings of Annie Besant and C. W. Leadbeater (later members of the Theosophical Society). They classified their knowledge into tidy schemes of 'planes' and 'subplanes' that people could easily understand. However, any scheme that seems simple is bound to be over-simplified, a rough general guide that omits the complexities, the exceptions and the principle that 'circumstances alter cases'. The Besant-Leadbeater scheme was in line with the over-confident trend of science in the late 19th and early 20th centuries. Scientists of that period firmly believed that everything was knowable and straightforward, and that in due time every branch of science would be neatly classified and finalised. Such complacency was later to be shattered by the theories of modern physics, as well as by the unsuspected complexities of Nature revealed by ever more powerful microscopes and telescopes, and other techniques for probing more deeply into her secrets.

Then the simple schemes had to be modified and made more flexible — and so it is with the system of 'planes' and 'subplanes', each of material more subtle than the last, which sought to explain the mysteries of the inner 'planes' in such clear-cut terms. The word *mysteries* was used advisedly. Each 'plane' must be experienced personally to be known, and each in turn is a revelation. The words used to describe it in the textbooks are then seen to be inadequate to express the wonder of the new experience. This always transcends common earthly happenings, which is all our everyday language is capable of describing.

Nevertheless, we do need some guidebook, something to tell us what we are to expect in the afterworld, however inadequate the information might be. Even to say that it will be more wonderful than we can imagine is some help and encouragement to go forward in faith and hope when the time comes. So I have to tell you that the scheme involves a number of 'planes', or to use a more modern term, levels of consciousness. The highest 'planes' are beyond human comprehension. For our purpose we shall seldom need to think about any but the three lowest 'planes', physical, emotional and mental and these can usually be regarded as subdivided into two or three distinguishable sections. These divisions assist our analytical understanding, but they are not sharp and definite, they overlap and mingle. It is not helpful to think of the multitude of 'subplanes' as resembling separate floors in a multi-storey building which can each be explored and lived in, without regard for all the others. The real situation is far more fluid and flexible than this.

Consider first the physical 'plane'. Here we can readily distinguish the three lowest 'subplanes', solid, liquid and gas. In the laboratory they can be studied separately. In Nature they mingle freely; in our bodies for example we have solid bone, liquid blood and other fluids and gas in our lungs. Yet even bones contain a little liquid; the blood has suspended solid red cells and dissolved oxygen gas; the air in the lungs has suspended droplets of liquid water and even minute solid particles if there is smoke in the air. Thus it is with the far more rarified substance of the higher 'planes'. But we have not yet finished with the physical level. Besides the familiar solid, liquid and gas it is said by some to include etheric sublevels. Few scientists would accept this claim, but I believe the many subatomic particles recognised by physicists belong in these etheric levels. Our bodies are said to include an invisible part, the 'etheric double', made up of this etheric matter. This body is concerned with vitality, the Life principle that distinguishes a living person from a corpse. It is also the link between the physical body and the psyche. At death this etheric body is put aside, along with the physical body, so it need not concern us much.

Next comes the emotional level or 'plane'. It is in the subtle body in which you find yourself in an out-of-the-body experience, or after death, already mentioned in earlier chapters. In many books this is referred to as the astral body. Its abode can conveniently be subdivided into three levels, easily distinguished after a little experience. It will normally contain material from all sublevels; but the proportions do vary with the person's temperament and manner of living, and later with his progress in the after-death life. Without this astral body in this life, we could not experience emotions. An important point to grasp is that astral matter from different levels varies in *quality*. That of the lower levels is relatively coarse and best suited to the expression of crude emotions like greed, lust, jealousy, hate and the animal passions. As we proceed to the higher levels the quality improves progressively, till astral matter of the highest level will only support the finest emotions, like unselfish love and aesthetic appreciation. Thus in practice there *is* some layering, arising from the principle that like attracts like and tends to exclude unlike. Some sensitives say that contact with the lowest astral sublevel is a most unpleasant experience; not only does it call up the foulest emotions, but it seems to convey a sensation of stickiness, like wading through treacle!

It must now be explained that life in the astral body after death is in several ways different from out-of-the-body experiences that may be had during earth life. First, when out of the body you are operating at an unusual level of consciousness but you are still a complete personality. Though you are temporarily separated from the etheric body of vitality and the familiar physical body you are still loosely attached to them and you can return to this safe refuge whenever you wish. You are functioning at the levels of emotion and thinking; but beyond the everyday concrete or lower mind stands the higher mind, the abode of the higher self, to which you can appeal for guidance. All these bodies (or available levels of consciousness) have their effect on the out-of-the-body experience. Moreover, if it happens to be an unpleasant one, if things get rough 'out there', you need only think of your ordinary body and you will instantly return to it, so ending the experience, except as a memory. But after death your escape routes are cut off. There is no longer a physical body into which you may safely awaken. The only way to escape from unpleasant experiences is to turn the mind to more pleasant thoughts like love for relations and friends. Then you will rise above the unpleasant events. Of course this escape route is also available in the out-of-the-body state, as an alternative to returning to the body.

The second way an out-of-the-body experience differs from death is that you are still loosely linked to the brain which acts to steady and concentrate one's thinking; after death you think with the mind alone, unsupported by any material structure, and it is going to be more difficult to concentrate. Thirdly, it is said that after death some changes may take place in the astral body; and fourthly you are liable to be cut off from the spiritual guidance of your higher self; but these are matters that will be easier to explain later on.

It would not be fair to lead you to think that life after death is bound to be a time of total bliss. It will eventually be so but first you have to earn your right. It is necessary for me to warn you about unpleasant events that may happen to us along the path to heavenly bliss. All of us must first pass through a sort of purgatory and some of us will go through a period of time resembling the Christian idea of hell. But this is not for eternity, though it may seem so at the time; also, as Christian doctrine explains truly, it is avoidable by right living on earth — though hardly by death-bed repentance. If hell must be experienced be sure that it is self-inflicted torture and that its nature is appropriate to the sins committed. But it is not punishment meted out by an angry God; it is expiation, which your own higher self recognises as necessary. The suffering you may experience has two main sources. First, if you have inflicted suffering on other persons, or even upon dumb animals, you must suffer the semblance of this wickedness yourself until you fully realise what you have done and you truly repent and firmly resolve never to behave in such ways again: as we shall see later, your resolve will be tested in due course. Moreover, the result is much the same whether you inflicted the suffering deliberately and callously, in retaliation for some wrong done to you, or merely thoughtlessly, though the time needed for expiation may be different. Also mental torture by unkind words calls down this kind of retribution just as much as physical torture.

Another source of unhappiness, already mentioned, is the yearning for earthly pleasures that can no longer be gratified in the absence of a physical body. Naturally this will most affect the sensuous person cut off in his prime by disease or accident. If you are spared to have a peaceful passing in old age, these earthly desires will probably have lost their attraction already and you will avoid much of this sorrow. The moral is obvious; in youth and middle age, enjoy your pleasures certainly, but cultivate a measure of detachment too. Do not become a slave to enjoyment of life's gifts; recognise that it is the lesser personality that they please, not your true inner self, which ought to be in charge of your life. Does this seem stern advice? Be assured, the more fully it is followed, the more reasonable and proper it comes to seem. When you falter and the disciplined life temporarily loses its attraction, you may reflect that steadfastness now will make it that much easier, later, to pass through the sorrows of the after-life.

I mentioned that certain changes in the astral body usually occur after death. In life it contains astral material from all sublevels blended together, so that the whole gamut of emotions can be experienced in quick succession, from the most base and selfish to the most lofty and unselfish. After death most people will undergo a rearrangement into shells, or coats of skin like an onion; the coarse material goes outside and the finer is buried within. A little reflection will reveal that this is a merciful provision however harsh it may seem at first, for the result is that you are compelled to face the lower reaches of the astral world first, in a short concentrated bout, so that you become free to rise to more pleasant levels without any fear of being drawn back to the gloom and pain. These will cease as soon as you have worn out your lust for life on earth. For this is the scheme of things and it is better to cooperate than to try unsuccessfully to return to life on earth which is now over. Sooner or later you will have to accept that life after death is a spiritual pilgrimage and that it will become increasingly understandable and enjoyable as you progress. After reading this you can face the early tribulation bravely, certain that great happiness lies ahead. As you begin to turn your back on earthly pleasures, this coarser astral material drops away, leaving the next layers exposed so that you can enjoy the finer emotions.

Perhaps I have dwelt too long on this phase, this period resembling hell. I have done so because I do not wish to incur the wrath or reproach of any reader. I do not want any of you to be able to complain that I painted a picture of joy and happiness, whereas you find yourself bewildered. By no means everyone has to experience this phase. We may all have to spend a little time in this condition, but you may pass it in a kind of sleep or dim awareness, instead of having to suffer the retribution and fear experienced by a really wicked person. You may have made amends for youthful indiscretions, you may have lived a good and abstemious life, been kind to relations and neighbours, giving honest labour for what money you received in your working life; perhaps you have tried to serve your community in some helpful way. If so, then you have nothing to fear. When you awake on the astral 'plane' you will be already in the Summerland the spiritualists describe, the middle region which I outlined at some length in Chapters IV and V. Here you can continue with the gentle pleasantries of life, in so far as they can be carried on in imagination, without a physical body. You may live at home with your former interests, hobbies and pastimes, meet friends and relations for conversation, games, music or whatever, tend your precious garden endlessly, travel or do anything else that takes your fancy. There are no duties you must perform, beyond any you impose upon yourself, no restrictions, no responsibilities. There need be no sadness — unless you actually enjoy feeling miserable!

But this perpetual round of enjoyment is just a way of passing time; it doesn't get you anywhere, and eventually it becomes boring, and you sense, or you are told, that there *is* somewhere better and beyond. By now you will have worn away and discarded further layers of your astral body, until you are left with only the finer grades, so you are automatically carried on to the highest astral realms. These are regions where more intellectual or religious pursuits are likely to be indulged in. Here are philosophers and scientists still busy in their libraries or laboratories continuing their earthly researches year after year. Here too are strongly religious people endlessly worshipping in their chosen church or temple. Here too musicians go on composing, poets and authors write more books. Perhaps none of these activities appeal to you. Then probably on earth you never built much of this finest astral material into your Body of Light, so this subtle body will soon wear away, leaving you free to go on still further, on to the mental 'plane'.

By now you have finally cast off all the impurities, the foolishness, the selfishness of your life on earth, and any lingering yearning for earth life. You have liberated yourself from this. All you can take on with you is the valuable essence of that life, recollections of the lessons it taught you, unselfish work for others you may have done, unselfish love relationships, and new powers you developed to enable you to do these things. You can now dwell on these precious fruits of your life and you can develop them in perfect peace and happiness.

But before we proceed in imagination on this further stage of our adventure, there are a few further pieces of information you may like to have. Possibly you are curious as to how long all this will take. This I cannot tell you, except in very general terms. It will take just as long as you personally need for what has to be accomplished, plus any additional time you choose to spend in this state, either because you like it or because you are scared of what lies ahead. A saintly person may only need a year or less; someone who has lived a full and eventful life, immersed in mundane concerns, may require fifty years or more to work free of all his involvements. But it will not matter to you anyhow, for you will be unable to gauge the passage of time and there is no reason why it should concern you. There is no night and day, no passing seasons in this world, nothing to draw your attention to the flow of time.

What I have to say next may seem strange and hard to understand. I mentioned earlier that in this transition state at the astral level we are cut off *from above* as well as from our former bodies below, and this must now be explained. You may or may not be clearly aware that the 'real you', the higher self, is distinct from the personality or lower self. It is this higher self that tries to guide your life through conscience and other intimations. But often you may ignore its messages, preferring to identify yourself with the personality, which has its own ideas and habits. So from weakness, habit or sheer wilfulness you may do things you know you really should not do, or from laziness you refrain from doing some good deed you feel you ought to do. So there is sometimes a struggle between these two parts of you, the higher self and the personality, and the conflict does not cease when the physical body is lost. The mental 'plane' is rather sharply divided into two parts. The higher self is at home in the higher part, corresponding to the higher or abstract mind, associated with Love-Wisdom, and Will. But the lower or concrete mind using the lower part of the mental 'plane' is part of the personality. At death then, the mind is split asunder. The higher self with the abstract mind withdraws and may be said to have gone to heaven already. It takes little further interest in what remains of the personality and leaves the feeling-emotional part to work out its time on the astral 'plane' along with the concrete mind. This may seem strange behaviour but actually it is another merciful provision, along with the changes in the astral body already mentioned. Cut adrift from the higher self, the personality remnant can do no further mischief; but neither can it be inspired to perform useful acts. In a sense it is an automaton, set to go on running along the same lines of activity and thought that it has developed in its life-time on earth. But its experiences will *seem* real and vivid enough. In other terms, this remnant still has its animal soul, but lacks the divine soul. If this is so, then it can be understood that the entity in this situation is unlikely to gain new information of any particular value, so it is not surprising that any communications it may succeed in making through mediums tend to be trivial. (See Chapter III). Moreover it is in a sense diminishing all the time, as it throws off successive layers of astral material. Eventually it is reduced to a shell, when all of permanent value is withdrawn, to be reunited with the higher self in Heaven. Yet before this shell disintegrates it could in some instances be drawn back towards earth, make rapport with a medium and regain spurious and temporary vitality from her. Any message might still be recognised as coming from the person claimed, but its necessarily banal content would distress the recipient. This is another reason which I could not explain in Chapter III for my disapproval of modern spiritualist practices.

During earth life we can be classified either as outward-looking extrovert types or as inward-looking introvert types. This applies only in a broad sense marking the general outlook of the individual. We are all forced to be extroverts part of the time when we have to cope with the unyielding physical world in our need to survive and ensure comfortable living. But after death the opposite holds; we are all obliged to become introverts, and those who tended in this direction on earth will adjust more easily. After death, outgoing activity ceases, except insofar as we continue it in imagination; our whole purpose becomes that of looking within, reflecting upon our past life, discovering what lessons it has taught us, what wisdom we have gleaned from it, and how we might do better if we had a chance of another life on earth. In other words, our experiences become wholly subjective, and the whole situation somewhat resembles a continuous dream which seems real enough at the time.

Scientists have found, as they probe ever deeper, that Nature's ways are exceedingly subtle and complex. The ways of human nature are perhaps the most complex of all because each person reacts differently to life's circumstances and experiences. Accordingly it should come as no surprise if I am obliged to admit that there are exceptions to almost every statement I have made in this chapter. They remain true in the majority of cases — but you might be one of the exceptions. Among these are people who die, not peacefully from old age or incurable diseases, but prematurely, such as by sudden death from accidents, heart attacks, drowning and so forth, or victims of suicide. For them the normal rearrangement of the astral body and the withdrawal of the higher self does not occur. They live on, often unaware for a time that they are dead, in exactly the same condition as in earth life except for the lack of a body. All the available literature agrees in regarding this as a tragic and deplorable condition and this belief is reflected also in the Christian prayer for preservation from sudden death. That is why I have called the normal changes merciful provisions to ease the soul through this transitional period. We are told moreover that suicides must remain in this state for the remainder of their originally allotted span of life, plus any additional time they need to effect their release from the astral levels. They remain vividly alive, frustrated by the absence of a body and often trapped in the lower levels of the astral 'plane', those most closely linked with earth. They can act quite deliberately for good or ill; because the higher self is not in full control they may act unwisely in their desperation.

In contrast with these are the advanced souls for whom the higher self is already firmly in charge. They have nothing to fear after death, no purgatory to endure, so they choose deliberately to prevent the rearrangement of the astral body and to maintain the link with the higher self; they prefer to spend their brief spell at this level in their usual fully conscious, responsible state. Another exception must be made for the philosophers, scientists and others I described earlier, busily continuing their work on the highest astral 'subplane'. Such work would necessarily involve the abstract mind, so we must assume either that the link was maintained, or that it was reestablished as they near the end of their stay on the astral 'plane'.

Thus we realise that conditions at this stage of the adventure are flexible: they are adapted to suit the particular needs of each individual, so that he can make the best use of his stay at these levels. The next important change now looms ahead, to be described in the following chapter.

CHAPTER VII

HEAVENLY BLISS

'So she died, and went to heaven', the fairy tale says. So she did and so shall we all, for this is a fairy tale that comes true for everyone. But it is not quite as simple and immediate as the tale implies. The natural home of the higher self is the realm which has been called by some the mental 'plane' and which I am going to refer to as heaven. The abrupt transition from earth life to heaven life would be too startling and indeed unwelcome to most of us. We need a period of preparation in an intermediate realm, and in this book the preparatory stage has taken us through several chapters. First I sought to convince sceptical and materialistically-minded western readers that heaven does exist, not so much as a place, but as a state of being. Then I tried to persuade them by the evidence available, that life after death need not be feared, that even its earlier stages could be an enjoyable adventure. At last we stand in imagination at the threshold of heaven. What will it be like?

First, the transition will not be sudden or unexpected. We have seen how people who die peacefully in old age often have intimations of the life ahead. They may remember, more clearly than in early life, visits to the next world in sleep or have conscious out-of-the-body experiences. Near the moment of death, we are told that they see what lies ahead to welcome them. It is indeed a common observation that the faces of the dead wear a happy peaceful expression. In just the same way, people facing this next stage of after-world existence have intimations of what is to be. They may have short periods in that pure and rarified realm, then retreat again to the familiar higher reaches of the intermediate realm. But at last they become ready and willing to pass on. By this stage they will have finally turned their backs on the attractions of life on earth, and their more refined counterparts at astral levels. They will have purified their souls from all the grosser emotions that cannot find expression in the heaven world.

Although this lies at what is known as the mental level, it would be a mistake to think of it as exclusively a world of thoughts — an impression one might form from reading some of the literature on the subject. Without doubt we must carry the loftier parts of our emotional nature into heaven, for unselfish love is depicted as an important feature of this state. So you may ask, as I had to ask myself, what then is the difference? At the higher levels of the astral realm we can exercise our purer emotions, and we have access to our lower mind. Now the emphasis changes; we dwell, as it were, in our mind and have access to appropriate emotions. So is this all that changes? No, it is not. The intermediate state we are preparing to leave (in imagination) was an artificial kind of existence in a way, because we were then separated from our true self; we were incomplete beings. Now at last, entering heaven, we come to our true home: we are reunited with our higher self, we become whole again. What is more, we have cast off all the tiresome aspects of our personality, those which held us back from realising God. Now we are liberated from all this; we can enter our real home at last in peace and in glory.

This new life is so different from life on earth that it is very hard to imagine. All the unpleasant features of life on earth are now forgotten; in heaven there is no sadness, no memories of unhappiness, no wickedness. We will recall only the happy, worthwhile events and the wisdom we have gained in the past life. Imagination is the key to this world, even more so than in the intermediate state. Whatever we imagine comes into existence instantly: we create our own heaven exactly the way we want it to be. If we happen to believe in the caricature of sitting on a cloud playing a harp, then that is just what we will do until we tire of it and think of something more interesting to do. If we are devout we may spend long ages worshipping in our particular church, temple or mosque, or finding God in nature. But if our idea of heaven is a more homely affair of happy family life, then that is the way it will be; family and friends, living or dead, will be around us in happy communion. How can this be, you may ask, when some of them are still alive on earth? I said that imagination was the key; these are mostly figures created by the imagination, though they will seem real enough and in a sense they *are* real because each of these images is a kind of outpost of the actual person, vivified by his own thoughts and feelings. In that world it *is* possible to be in many places at once even without being consciously aware of it. But if the thought or love of another person is strong enough, and he or she is also in heaven, then we may actually meet and commune consciously. If this happens it is a communion more intimate, complete and satisfying than anything we experience on earth; we become, as it were, a single united thought.

If however our ideas of heaven are of a more intellectual or artistic turn, then again we will not be disappointed. We may spend our time exchanging ideas with others of similar interests, learning or teaching, as preferred. It will not be like a debate or discussion on earth, delayed and hampered by inadequate words, liable to be misunderstood. Communication is by a sort of telepathy of the ideas, the thoughts themselves, not their clumsy expression in words. If we are planners or creative thinkers, we may go on making splendid plans, devising new civilisations where all people are wise and happy. Or we may create wonderful new philosophies that will guide men towards wisdom and harmony. There is no limit to what we may do, allowing the imagination free rein. Yet again we may listen to glorious music, beyond anything that could be produced by earthly instruments, or even compose such music. We may watch, or even write, great plays or operas. And so on; whatever we do or enjoy is what we *want* to do; a continuation of the activities we loved best on earth, carried on without hindrance or frustation by earthly limitations.

Does all this sound too good to be true? It will not be. The reality will be far more wonderful than can possibly be imagined. Yet in another sense, yes, it is too good to be true. Let me try to explain this contradiction. When, in earth life, we gain a glimpse of the next (astral) world, we discover that it is beyond description. It has its unique qualities: in many ways it is quite different from anything on earth, and quite wonderful. But if we could rise in full consciousness still higher to the next major level, which is this heaven world we are talking about, exactly the same thing would happen to us. The experience would be a totally new revelation, of a state of being more glorious, more free, more uplifting than we could ever have imagined. For this mental level too has its own uniqueness, its own special, unforeseeable qualities. But in practice very few of us are able to make this transition while on earth, so we have no experience of this strange new world, we do not know how to cope with it, how to find our way about. When we reach this sphere at last, after death, we are like new-born babies on earth. But this is an imperfect analogy; the baby soon learns. When we reach heaven, however, it seems that we are not able to experiment and learn as a baby does; we can only make use of any experiences we gained while we still retained a physical body and a brain to help us sort out these new experiences. But most of us in earth life are not yet ready for such exalted experiences: we still have more down-to-earth affairs to occupy us fully. So we arrive in heaven unprepared and the most we can accomplish is to create our own bit of heaven around us and dwell in it subjectively in a kind of dream state. After reading one rather biased account of life after death, I jotted down in my notebook 'a thousand years of daydreaming', as a summing-up of the heaven period. Later I realised that this was a travesty. To a yogi of long experience, who had gained the freedom of this mental level, the situation of most inhabitants of heaven would indeed appear like this, but to each of these individuals it does not seem like this at all. For each of them it is a full and blissfully happy life in perfect freedom and that is what matters — our own feelings about the situation. I have already said that conditions in this world are really indescribable, and all this may not make much sense at present. Rest assured; we will enjoy it enormously when we get there.

However, let me try again to make matters a little clearer. Another account of life after death suggests that it is a cruel delusion to regard life in heaven as mere dreaming and illusion. In physical life, it is pointed out, many of us are highly self-centred and operate only within a limited region. We are limited, not merely in regard to our places of abode and work, but in our thoughts and ideas, living in our own little world of prejudice, habit and convention. So if we do the same in heaven, it is no surprise. Yet it is not the same; in heaven we are that much nearer to the source of our being, so we see and think more truly. Nor must we suppose that this long stay in heaven is useless. On the contrary it is purposeful and indeed it serves a dual purpose. On the one hand we 'reap our reward in heaven' for the trials and tribulations of a strenuous life well-spent on earth; we enjoy peaceful relaxation and reflect upon our achievements in the past life. On the other hand those reflections have positive value; we have ample leisure to digest our many experiences and draw out from them their virtue. So we transmute experience into wisdom, and unrealised aspirations into the faculty to bring them to fruition, should we be given another chance. It takes time and freedom from commitments to do all this; power is granted freely, to the limit of our needs and abilities. So how long do we need? Some say that life in heaven is eternal. I have already hinted at a period of a thousand years, and it is said that people who have lived very full lives may indeed need even longer. For most of us a sojourn of a few hundred years seems probable but we really do not know for certain and there are likely to be big variations from one person to another. Life on earth seems to be moving faster and faster, as science and civilisation bring rapid changes in our way of living, and also rapid increases in world population. So, some people suggest, the heaven life may be speeded up also. In the absence of any events to mark the passage of time, the experience will be timeless and time, as we know it, will not be relevant.

Like the astral 'plane' described in the previous chapter, the mental 'plane' is also said to be subdivided into a number of sub-levels, differing in the quality of their rarified mental substance. However a more important division into two parts has already been mentioned. The lower half is associated with concrete thoughts and is part of the realm of the personality, the lower self, while the upper part is concerned with abstract thinking and is the abode of the higher self. It would appear that the soul in heaven is not obliged to experience conditions in the various sublevels one by one; instead it is drawn to the level appropriate to its mood and inclinations.

So what are the options? It is said that the first level is the natural abode of those many people who asked for nothing more on earth than an affectionate family life and a circle of good friends. If this is your inclination, then you will be well pleased;

your family and friends are there with you whenever you want them. There is no need to take journeys, as on earth, to visit sons and daughters who have settled far from the family home. Just think of them and they are with you in loving relationship, along with sons- and daughters-in-law and your grandchildren, some perhaps that you were never able to meet on earth, or possibly born after you died. Unrelated friends, and their families too, all are available on demand. Now what about the family pets you may ask; you missed them when they died, or perhaps you went first and left them to mourn for you. Yes, you can have them with you. It will be in imagination, actually, though it will seem so real that you will not realise this. The animals may not be able to reach your level, but the image you create, by thinking of them, will be animated and vivified by the affection they bore you on earth. As I explained earlier, it is the same with friends. You cannot claim them exclusively; that would be selfish and unfair because they have other friends. In heaven there is no place for selfishness or injustice, but it will not arise, for you can see as much of them as you wish. If you tire of their company you have only to withdraw your attention and they will fade away.

The second level is for those who looked beyond family and friends for guidance, who worshipped some religious figure such as Jesus, Buddha or Mahommed, or some particular Saint, or one of the Hindu Gods, according to their accepted religious sect. If you went to church only on Sundays, then doubtless you could continue something like this practice, spending most of your time in the first level and visiting the second whenever you felt the religious urge. There is no longer any compulsion from local custom or a sense of duty; you may do exactly as you now feel inclined.

The third level is largely for those whose religious devotions were of a less personal nature, those who worshipped God, or Cosmic Intelligence, or Brahma in some more abstract fashion. Also for those who saw the Divine Principle enshrined in men and women, and who strove actively to lead helpful lives.

The fourth level is for those of more intellectual bent, who took this abstraction a stage further, seeking to work for the good of humanity rather than for particular individuals or groups of people. Here are people who made an unselfish pursuit of spiritual knowledge, the more inspired philosophers and scientists, great writers, actors, musicians, artists and so forth; people who tried to convey something of their own spiritual understanding to others.

There is no need for more precise details and examples. By the time we reach the heaven world, we will be thoroughly used to life without a physical body. Once again it must be emphasised that there is not the slightest need to fear this new transition, this next stage of our adventure. It will certainly be more congenial than the intermediate one, and we will find our way without any guidance. Without conscious effort, we will find ourselves in the conditions best suited to our needs of the moment — and in others equally suitable when the needs change.

There are still higher reaches of this world, but the only people who would feel at home there are those who were thoroughly familiar with abstract spiritual thinking on earth. I refer to people who understood the full purpose of life and who were prepared to sacrifice worldly enjoyments to cooperate with these spiritual ends. As I shall explain in my remaining chapters, there are indeed many souls in the higher levels. Most of them are not ready for full waking consciousness in that holy realm, so they spend the time in a sort of sleep filled with blissful dreams — a condition that cannot be described without leading to misconceptions. However it is said that everyone has at least a conscious glimpse of the higher level, where we stand naked, revealed as our true, eternal, higher selves, finally bereft of the last remnants of the personality we have assumed since birth.

In conclusion, let me re-emphasise that the key to a long and happy stay in heaven is the way we think and behave on earth. Every kind thought counts and every good deed done spontaneously, without any selfish hope of repayment.

CHAPTER VIII

UNFINISHED BUSINESS

There is a grand purpose that runs through our life on earth and is brought to some sort of conclusion in the intermediate and heaven worlds. I have hinted before that during this last period we extract a kind of essence of abiding value from life's experiences; we transmute them into wisdom and into capacity for purposeful action. We ponder upon our successes, upon our failures and idealistic schemes that came to nothing, wondering how we might bring them to fruition if we had another chance. But what use is that if we are never to be given another chance, if we have lived our one life, and this is the end?

I have indicated also that heaven life is not eternal, that eventually we tire of this long introspective dream-like condition and we would be glad to bring it to a close because once again we long for action. We have been thinking fine thoughts, devising wonderful schemes that need to be tried out on earth. So we pray: 'Lord, let me go back, let me have that second chance and this time I will try to do better.' I have said there can be no sadness in heaven, so our prayers must be answered. We are living by now wholly in our higher selves, the personality of the past life is finally dissolved and we have drawn into ourselves whatever of value it held for us. So the decision is made by the real inner self; we really and truly want, more than anything else, to come back to the excitement and challenge of earth-life, to prove to ourselves that we can make a success of it. So be it!

Every day on earth hundreds of thousands of souls are born as living babies; so surely a suitable birth can be arranged. Why should all these souls be newly created when there are many waiting to be reborn? Perhaps you have scoffed at the idea of reincarnation or not known what to think about it. But, explained in this fashion, does it not seem entirely reasonable, even inevitable? It is not such an uncommon idea these days; more than half the population of the world do believe in reincarnation as an article of faith. You may well ask: if that is so, why is the idea not taught in the other great religions, and particularly in Christianity? There are several answers to this question. One is that the people to whom Jesus was preaching were already familiar with the idea, so there was no need to emphasise it further. There are several passages in the New Testament that seem to imply reincarnation. Some scholars say there were more such texts in the early versions but they were deliberately removed from later translations by order of one of the Church Councils. There could be tactical reasons on the part of Jesus Himself, or of the Church, for avoiding anything more than minimal references to reincarnation. People who are already indolent by nature may be persuaded by belief in reincarnation, not merely to put off till tomorrow what they should do today, but to put off till the next life what they should accomplish in this one. If, on the contrary, they believe they have only this one life they may be inspired to make the most of it; thus they may put forth more effort, gain more experience and altogether make more progress in this life. Whether or not it was predestined, we can see that this attitude, over the last 2000 years nearly, has indeed led to a vast speeding up in the advance of civilisation. We tend to complain that undue attention has been paid to material progress, to comfortable living and amenities beyond our true needs. But also there have been vast improvements in education and cultural pursuits; at least we now have leisure and opportunity to lead a spiritual life if we wish to do so.

The other big question people ask is this: if I have lived on earth before, why do I not remember it? To this problem also there are satisfactory answers. One is that some people *do* remember past lives, occasionally in considerable detail, more often just in flashes. But the majority certainly do not remember and this is not really surprising. After all, how much do you remember of the earlier part of your present life? Today you are reading this book. Stop a moment and think back. What exactly were you doing on the same day 30 years ago, even ten years? You cannot remember unless it was a red-letter day, you have an exceptional memory or you have looked at an old diary. You do not have diaries for previous incarnations! You may remain unconvinced by these arguments. There *are* days you can recall vividly, so why not the highlights of a previous incarnation, even if it was hundreds of years ago?

A more fundamental argument concerns the problem of who or what reincarnates. I have already stressed that the 'YOU' that begs permissions to return is your true inner self, the real abiding 'YOU', the higher self. But how often in ordinary daily life, do you think of yourself in this way? The normal 'you' is the lesser self, the personality, urgently concerned with day-to-day problems, desires, food, drink and enjoyment — everything that makes up daily life. It is only when you are faced with big decisions, or on Sundays perhaps, that you put aside worldly thoughts and become, briefly, your true spiritual self. So this personality, which is 'you' for perhaps 95% of the time, this 'you' that remembers the incidents of your life, this is the entity that slowly dissolves away during the afterlife. *It* does not reincarnate. By the end of life in heaven, most of the details of its earth-life are already set aside and forgotten. How could they be recalled, in the midst of a future busy life on earth?

What *does* reincarnate is that pure essence, that everlasting 'YOU' which occupies perhaps 5% of your time during earth-life. This 'YOU' is not much concerned with events, with incidents in their own right; these can be left to the personality, whose proper concern they are. The inner 'YOU' remembers not so much the event itself, as the abiding effect it had on your character. Most of life's little events are too trivial to come into this category at all. So the reincarnating entity has already lost interest in the past life, except for its really meaningful events. Attention is turned to the future life, and when it comes about 'YOU' will properly be interested in that *new* present time, in bringing to fruition all the noble plans 'YOU' dimly remember working out in heaven. Why bother about the past that was lived by *another personality,* for that is the true situation. In the new life, a new personality is gradually built up as childhood progresses into youth and into adult life.

A few people *do* remember their previous life on earth in

considerable detail, so these exceptional cases must be considered further. They seem to occur, or to be noticed, mainly in India and other eastern countries, and the memories start in early childhood as soon as the children can speak about them. In these countries where the idea of reincarnation is accepted such memories would not seem very strange and the parents would listen. Similar cases do occur in western countries also, but parents tend to ridicule the child's stories of a previous life so that, with the precocious wisdom of children, the child stops talking of his memories and soon loses them.

Dr. Ian Stevenson has collected scores of cases; in each one rebirth had occurred within a few years of the end of a short previous life so that it was not too difficult to discover from the child's descriptions where it had lived previously. Then with Dr. Stevenson's help the child would be taken to the place, which it would recognise, and would find its way to the previous home, often only a few miles from the present one. Still-living relatives would then be able to confirm the child's account of the former life, sometimes given with great and intimate detail. Dr. Stevenson applied his skill as a psychologist to analyse these cases, to try to find more ordinary explanations, but he was driven to the conclusion that reincarnation was indeed the most acceptable one to explain all the evidence.

I said in another connection that every rule had its exceptions, and I believe these cases do represent exceptions to the usual course of events. Since, for some reasons we cannot fathom, these individuals incarnated again so soon after death, there would not have been time for their former personalities to disintegrate in the gradual manner I have described. Accordingly, they were probably born again retaining their former personality complete with its memories. They obviously had new physical bodies, but it seems likely that their emotional and mental aspects were carried over from a previous life. So it was almost as if they were recalling events of their *present* life; that is to say, they had continuity of memory. But since this does not happen for most of us, it is not surprising that we have few, if any, recollections of previous lives. Sometimes however, we do recover a flash, as when we seem to recognise a person or a place we have not encountered before in this life.

I have entitled this chapter, 'Unfinished Business'. Rebirth is by no means a random affair. Like everything else it is governed by rules, by natural law, though this one is not yet recognised by science. In the east it is known as the law of karma, the law that arranges matters so that you shall reap what you have sown, good or ill. If the reaping was not possible in the same life as the sowing, then it is deferred till the next or a later life. What happens to us is in full accordance with the working of this law, though often we do not recognise this. The present personality does not know, and this may seem unfair; but the higher self knows. The trouble is that most of us do not listen to its inner voice: we are too busy letting our personality have its own way. Let me pause for a moment to extend a point made in the previous chapter. I said that at the end of our heaven life we should rise up into a higher level and should live for a while as our true selves. It is said that in that period we are granted another review of the past life, such as the one we had immediately after death. But this time we are shown how this life keys in with previous lives, and we are granted a brief preview of the life to come on earth. Thus we see how we can pick up the threads again, repay debts and receive favours owed to us, how we could mould this new incarnation to serve our inner purposes. In this higher level there are many souls waiting, dreaming away happily; they are waiting for a *suitable* birth. We must wait to be born into the right family — people we have met before, perhaps — and into a suitable environment for the next stage of our long journey. Thus is karma, or divine justice, in operation. It is not punishment for wrongdoing, but rather restitution, making amends. Nor, on the credit side, are there material rewards for good deeds though these may come, but we are offered opportunities to do better, to try to work out in earth-life the wonderful schemes thought of in heaven. Bonds of loving relationships are likely to be renewed, perhaps in new forms. But hatred also forges bonds; it must be expunged eventually by both parties settling their differences amicably, even if they no longer remember what the original quarrel was about.

The *Tibetan Book of the Dead,* referred to in Chapter VI, besides being a funeral service, is a book of guidance for the living. So too this little book is a guide not only for the dying but for those in the full spate of life. Worthy living will bring its ample rewards in heaven. Worthy use of the stay in heaven will bring fuller opportunities in the next life. So we can glimpse an overall purpose in this cycle of birth, life, death, heaven and rebirth. Each stage is a kind of rehearsal for the next; each full cycle a rehearsal for the overall scheme. This has perhaps never been summed up better than by Jesus, when He said to His disciples, 'Be ye therefore perfect, even as your Father in Heaven is perfect.' That is the long-term objective, and for most of us it will require many more lives.

EPILOGUE

PREPARATIONS FOR THE JOURNEY

This book is about a journey, one that all of us must take as life closes. Let us look forward to it as 'Our Last Adventure', the greatest we have ever undertaken. What a pity it would be to spoil it by unreasonable fear and foreboding. This journey will be a new experience for us, and it is a perfectly natural one. It has already been taken by countless millions of other people, but any journey needs planning and preparation. This Epilogue follows on logically from the later chapters of our 'commonsense guide'. It is a short summary, setting out the important points that should make our journey a success.

It has become fairly clear to me what I should try to do in preparation for my own journey, but readers may welcome some guidance. How should we get ready for this new kind of journey? It is unlike any other we have undertaken, so the analogy of a holiday trip is no longer much help. We need to consider again the purpose of life on earth and, more importantly, to think about the purpose of life 'on the other side'. For it is this wonderful new life that we shall be entering into when the actual journey through the portal of death is undertaken.

So, as we turn our thoughts inwards and reflect upon the meaning of life, we need to think about life in heaven as well as our earthly existence, to make the cycle complete. For me, even that is not enough. I believe that this cycle is just one of many. We need to think about the entire series, future as well as past. If you have been brought up in the Christian faith, and you cannot bring yourself to believe in reincarnation, this makes no difference. The conclusion is still the same. If you feel certain that you have only one life, then your need is that much more urgent to complete what has to be done in your remaining days here. The message is 'Perfection'. Does this seem an impossible task? Not really, for there are various stages on the way to perfection. It is like a staircase — having reached one landing, which once seemed unattainable, we shall see that there are still more stairs to be climbed.

It may be helpful to think of the key to our quest as the idea of *letting go.* What could be simpler? Carrying it through may not be so easy, but it really does seem worth trying.

You know the familiar saying that 'you cannot take it with you'. This refers, of course, to material possessions. You will be forcibly separated from these by death, however much you dislike the idea. But you can let go of your *attachment* to material things, in imagination, here and now. This will help to lessen the sense of loss when you are 'on the other side'. You might even decide that you no longer need some of them. It would certainly help the people you leave behind if you have tidied and sorted your possessions and perhaps disposed of some of them. You may like to re-make your will.

We *can* take with us our thoughts and feelings, prejudices and habits, faults and virtues. But do we want to? We will retain those parts of our character that are good and positive but why not learn *to let go* the negative parts now? We will become happier, more serene and it could speed our passage to the heaven world.

There is a way of *letting go* that does not call for too much effort — a simple form of meditation. Try it when you are sitting quietly, resting, in bed, just before going to sleep or if you wake early. Encourage your thoughts and feelings to turn towards your higher self. If you are used to praying, offer your devotion to God. Try to think of heavenly things. There is no need to force yourself. If mundane thoughts creep in, do not let them worry you; just recognise them as such and quietly go back to your higher thoughts. This simple meditation is suitable for restless western minds particularly in one's later years. It can be done any moment when you are quiet and after a time you will find that peace and tranquility will replace anxiety.

Let us look at our faults and shortcomings. Let us recognise and admit them. If we want to change them, it is better to replace them than to fight them. *Let go* the old ways, try new ones and see what happens. For example, you may have a hasty temper. Be aware of it as a fault you would gladly be rid of. Think positively of its opposite, of peace, harmony and serenity. When you are provoked, your temper will flare again, automatically. But it will slowly become alien to you; it will not last so long next time. Eventually you will 'nip it in the bud'; you will catch yourself feeling anger, then with a smile you will let go of it, and feel serene again. Try it please, and see its effectiveness for yourself.

This is somewhat akin to the bio-feedback technique used in some forms of psycho-therapy. Under skilled guidance, you teach yourself how to control some normally unconscious bodily behaviour. An appropriate meter is connected to you, and whenever you do the 'right' thing, its pointer moves. You gradually discover that what you are doing makes you feel better; it may control asthmatic attacks, for example, or reduce the severity of migraine. So you keep on doing it, and you may make a full recovery. In this way you can cure physical ailments. By the simple meditation I have just recommended, but without any instrument to help, you may succeed in curing 'moral ailments'.

In the last chapter, I wrote about the personality, 'you', the lesser self that demands perhaps 95% of your attention. This leaves only 5% for TRUE LIVING as 'YOU' the Greater Self. What I am recommending now is the attempt to raise this proportion greatly. How wonderful it would be to reverse the figures. Then for 95% of the time YOU would be REALLY ALIVE in your true realm, with only the remaining 5% spent in attending to the needs of the body. This is a counsel of perfection, of course. Very few of us could hope to achieve it in just a few years. If we did, we should become truly spiritual individuals, living saints, and a shining example to our fellows. But even the smallest move in this direction is so very worthwhile. I sincerely believe that this is the way we are destined to go in the after-life. I also believe that life in heaven is a rehearsal for our next life on earth. Furthermore, a whole cycle of birth, life on earth, death, life in heaven, and rebirth, is in turn a rehearsal for our full cycle of lives. A description of this complete cycle, taken from ancient scriptures, could fill a whole book.

Very briefly, Life is said to descend from virtual unconsciousness in Paradise or the spiritual worlds, through levels increasing in 'density' and differentiation, till it reaches the physical plane. Passing through the lower kingdoms of nature, living forms become increasingly different from one another, as species multiply. Eventually the stream of Life subdivides into individual units at the stage of primitive man. At each stage, the degree of consciousness increases, till at the human stage it blossoms into responsible self-consciousness. Humanity reaches the stage of deepest immersion into physical life. But, so we are told, around the present period in historical time we are beginning to emerge from this 'fall' and to grasp the idea that we must start the long ascent back to Paradise. When this quest is completed by any individual he will re-enter Paradise as a fully self-conscious Spiritual Being, gathering up all he has learned in this cycle of lives. This is the Path of Perfection that we must all follow in the fulness of time, when we make ourselves ready. But in the heaven stages we can only dream about it, so to speak. We are limited to contemplation, to devising brave schemes in abstract thought. Then, time and time again, we must face the test by reincarnating. How well can we work out these schemes in life on earth, against the stubborness of physical matter itself? We shall also have to contend with the pressure to conform with the relatively trivial modes of life around us. That is the difficult part, the crucial test of our ideals.

This is the way we are all destined to go, so why postpone the endeavour? Whatever we can achieve in the rest of our present life on earth is solid gain. This is not ephemeral dreaming, but actual achievement, here and now, in character-building. What we succeed in doing now will endure. It will give us added strength to 'fight the good fight' in the next life on earth.

In conclusion, something further needs to be said. When all the evidence has been presented, when the pleadings have been made, another element is required for completeness. It cannot be presented by the writer; it must be provided individually by each reader. This element is *faith.* It is no use hoping that some day scientists will provide convincing proof of these matters. They are beyond even the terms of reference of science. So in the end they must be taken on trust, and the basis of that trust is faith. Perhaps I should try to describe what I mean by faith. I sincerely believe that what I have written in this book is true. There may be errors of detail; there are bound to be ill-chosen words that may convey wrong or inadequate impressions — that is inherent in the very nature of language. But in essence I have faith that it is right. This *faith* is based on direct personal experience confirmed by what I have been told or have read, that goes beyond it. Faith is much more than just intellectual acceptance of evidence. It is an unassailable conviction that the Universe, our planet earth and our lives are *not* chaotic as they appear to some eyes. On the contrary, all these matters are ordered by some superhuman Cosmic Intelligence, some Power, some Being perhaps, working out a Supreme Purpose in ways that are inscrutable and puzzling. My faith includes the certainty that there is some Source from which we can derive individual inspiration and guidance. Such *faith* has been affirmed by all mystics throughout the ages.

Perhaps you already have this kind of *faith*. If so, I hope it has been reinforced by this book. If you were sceptical or agnostic, I shall dare to hope that the book has led you towards the awakening of some faith and that it will become stronger as you ponder these ideas.

SUGGESTED FURTHER READING

The first six books are modern accounts of life after death, similar in objectives to *Our Last Adventure.* They are broadly in agreement, but differ in details owing to the difficulty of conveying other-worldly experiences in ordinary language.

L. J. Bendit *The Mirror of Life and Death,* Quest Books, Wheaton, Illinois.
Draws upon the knowledge of the psychic Phoebe Payne (who became the author's second wife). Rather academic in style.

Geoffrey Farthing *When We Die,* Theosophical Publishing House, London, 1968.

Geoffrey Farthing *Exploring the Great Beyond,* Quest Books, Wheaton, Illinois, 1978.
The two books by this author are based largely upon H. P. Blavatsky's monumental work *The Secret Doctrine* and *The Mahatma Letters to A. P. Sinnett.*

Dion Fortune *Through the Gates of Death,* Inner life Publishing Society, London.
Written by a trained psychic. Full of sound advice, for example the effects upon the departed of excessive mourning and attempts to communicate. Short and easy to read.

Charles Hampton *The Transition Called Death,* Quest Books, Wheaton, Illinois, 1943.
'It is we in the physical world who are really the dead because we are in a world of only three dimensions.' Very informative and easy to read.

John White *A Practical Guide to Death and Dying,* Quest Books, Wheaton, Illinois, 1980.
Intended as a guide-book, but written in a more impersonal style than *Our Last Adventure.*

Leonora Eyles Strength of the Spirit, Constable, London, 1930.
A thought-provoking novel. The author displays insight into both inner and outer worlds.

Joan Grant *Life as Carola,* Methuen, London, 1939.
A gripping novel based on memories of a former incarnation.
Geoffrey Hodson *Reincarnation, Fact or Fallacy,* Theosophical Publishing House, India, 1967.
Another work by a trained psychic.
Elisabeth Kubler-Ross *On Death and Dying,* Souvenir Press, London, 1982.
Near-death experiences.
Mark Pelgrin *And a Time to Die,* Quest Books, Wheaton, Illinois, 1962.
A man dying of cancer struggles right to the end to discover his inner self. Intensely personal, moving and inspiring.

The following nine books are specially recommended in relation to the Epilogue. Most of them were written by scientists (some of them world-famous) during the last ten years.

Fritjop Capra *The Tao of Physics,* Fontana/Collins, London, 1976.
A popular book showing that the theories of modern physics on matter and energy are close to the teachings of ancient oriental scriptures.
Fritjop Capra *The Turning Point,* Wildwood House, London, 1982.
Extension of the previous book to human affairs.
Robert Ellwood *Finding the Quiet Mind,* Quest Books, Wheaton, Illinois, 1983.
Robert Ellwood *Finding Deep Joy,* Quest Books, Wheaton, Illinois, 1984.
These two books guide the reader step by comfortable step towards a realisation of inner harmony.
Fred Hoyle *The Intelligent Universe; A New View of Creation and Evolution,* Michael Joseph, 1984.
The famous astronomer comes round to the view that the Universe must have been created by intelligent design. A copiously illustrated popular account.
Jim Lovelock *Gaia: A New Look at Life on Earth,* Oxford University Press, 1979.
Persuasive hypothesis that conditions favourable to life are maintained in the environment by the 'Earth Goddess' Gaia.

Peter Russell *The Awakening Earth,* Routledge and Kegan Paul, London, 1982.
A splendidly optismistic view of the future when world government is run by practical mystics.
Rupert Sheldrake *A New Science of Life. The Hypothesis of Formative Causation,* Blond and Briggs, London, 1981.
The behaviour of plants, animals and men is influenced by what has happened in the past — a pervasive telepathic 'memory of nature'.
E. Lester Smith *Intelligence Came First,* Quest Books, Wheaton, Illinois, 1975.
By the present author. Intelligence, consciousness and life are the primal realities, not matter.

The following two books came to my attention after I had finished my book.
Avis D. Carlson *In the Fullness of Time,* J. Curley, Massachusetts, 1979.
A look at retirement and old age. Large print edition, available through Public Libraries.
Michael Marsh *A Matter of Personal Survival, Life after Death,* Quest Books, Wheaton, Illinois, 1985.
A study of survival with convincing and plausible evidence.

REFERENCES

For completeness, and for any who wish to pursue the topic in greater detail, books mentioned in the text and the sources of quotations, are listed below.

CHAPTER I

Wilder Penfield *The Mysteries of the Mind,* Princeton University Press, 1975.
Written by a brain surgeon, whose own research drove him to the conviction that the mind is separate from, and superior to, the brain.

Karl Popper and John Eccles *The Self and its Brain,* Springer-Verlag, Berlin, London, New York, 1977.
A big scholarly volume written jointly by a philosopher and a neuro-physiologist, both celebrated men.

Celia Green *Lucid Dreams,* Institute of Psychophysical Research, Oxford, 1968.
Readable little book, based partly on unpublished cases, partly on quotations from other books.

Celia Green *Out-of-the body Experiences,* Institute for Psychophysical Research, Oxford, 1968.
Based wholly on new cases, received in response to a broadcast appeal.

P. D. Ouspensky *On the Study of Dreams and on Hypnotism,* Chapter VII in *A New Model of the Universe,* Routledge and Kegan Paul, London, 1960.

O. Fox *Astral Projection,* University Books, Inc., New York, 1962.

S. J. Muldoon and H. Carrington *The Projection of the Astral Body,* Rider, London, 1963.
Muldoon's own interesting personal experiences, edited by the senior author.

R. A. Monroe *Journeys out of the Body,* Doubleday, London, 1971.
The author's own experiences, much more diverse and extensive than those of Muldoon.

J. H. M. Whiteman *The Mystical Life,* Faber and Faber, London, 1961.

Analysis of the mystical experience and lesser 'separations' illustrated by selections from several thousand personal episodes.

R. Crookall *What Happens When You Die,* Colin Smythe, London, 1978.
Somewhat presumptive deductions on life after death from analysis of out-of-the-body experiences.

CHAPTER II

R. A. Moody *Life After Life,* Bantam/Mocking Bird Books, London, 1977.
Original research on near-death experiences, and thoughtful analysis of the findings. Highly convincing.

R. A. Moody *Reflections on Life After Life,* Bantam/Mocking Bird Books, London, 1977.
Further research in the same field as the previous book.

R. Crookall *The Study and Practice of Astral Projection,* Dynamic Press, London, 1978 (latest edition).
An earlier book which also recounts some of the experiences described by Moody.

CHAPTER III

W. J. Crawford *Experiments in Physical Science,* J. M. Watkins, London, 1919: E. P. Dutton, New York, 1919.
Down-to-earth study by an engineer of ectoplasmic emanations from a medium used in table-lifting, etc. Published more than 60 years ago, but still valid. Should be read by sceptics.

J. G. Fuller *The Ghost of Flight 401,* Corgi Books, London, 1979.
Reads like a thriller, but it is a true story.

CHAPTER VI

Joan Grant *Winged Pharaoh,* Arthur Barker, London, 1937.
A thrilling novel, probably based on memories of a previous incarnation.

W. Y. Evans-Wentz *The Tibetan Book of the Dead,* Oxford University Press, 1957.
A long scholarly work. Interesting, but not light reading.

I. K. Taimni *The Science of Yoga,* Quest Books, Wheaton, Illinois, 1975.
One of the best of the many translations of, and commentaries on, the *Yoga Sutras of Patanjali.*

CHAPTER VIII

Ian Stevenson *Cases of the Reincarnation Type,* University Press of Virginia 1980, III.
Case studies by Dr. Stevenson.